AF819140

Optimizing Decision-Making in Engineering

This comprehensive book explores the evolving landscape of decision-making processes for engineering and IT professionals in increasingly heterogeneous technological environments. It addresses critical gaps in current educational materials by examining how technological advancements, situational awareness challenges, human–machine intelligence integration, and virtual uncertainties impact modern decision-making frameworks. The book provides a complete journey map of decision-making that connects theoretical foundations with practical applications, offering readers a structured approach to navigate complex adaptive systems in our data-driven society.

Optimizing Decision-Making in Engineering: A Journey Mapping Perspective includes an in-depth exploration of cognitive and emotional engagements in decision-making, an analysis of how advanced design functionalities influence cognitive biases and heuristics, and an examination of decision-making under virtual conditions where physical touch is absent. It presents analytical tools such as decision trees, Bayesian networks, and multi-criteria decision-making models (AHP, TOPSIS) while incorporating emerging technologies like artificial intelligence, virtual/augmented reality, IoT, and big data analytics. Real-world scenarios, use cases, and strategic frameworks like the Cynefin framework and Roger's model provide a practical context for theoretical concepts.

Designed for technology professionals, including CTOs, IT project managers, systems architects, UX directors, and engineering team leads, this resource equips product managers, business analysts, solution architects, and digital transformation consultants with enhanced decision-making capabilities for complex environments. This comprehensive guide delivers particular value for specialized roles such as AI ethics officers, human factors engineers, and digital experience strategists, providing frameworks applicable across autonomous systems, healthcare technology, immersive experiences, retail analytics, and advanced manufacturing operations.

Optimizing Decision-Making in Engineering

A Journey Mapping Perspective

Zian Shah Kabir

CRC Press
Taylor & Francis Group
Boca Raton London New York

CRC Press is an imprint of the
Taylor & Francis Group, an **informa** business

Designed cover image: Shutterstock

First edition published 2027
by CRC Press
2385 NW Executive Center Drive, Suite 320, Boca Raton FL 33431

and by CRC Press
4 Park Square, Milton Park, Abingdon, Oxon, OX14 4RN

CRC Press is an imprint of Taylor & Francis Group, LLC

© 2027 Zian Shah Kabir

Reasonable efforts have been made to publish reliable data and information, but the author and publisher cannot assume responsibility for the validity of all materials or the consequences of their use. The authors and publishers have attempted to trace the copyright holders of all material reproduced in this publication and apologize to copyright holders if permission to publish in this form has not been obtained. If any copyright material has not been acknowledged please write and let us know so we may rectify in any future reprint.

Except as permitted under U.S. Copyright Law, no part of this book may be reprinted, reproduced, transmitted, or utilized in any form by any electronic, mechanical, or other means, now known or hereafter invented, including photocopying, microfilming, and recording, or in any information storage or retrieval system, without written permission from the publishers.

For permission to photocopy or use material electronically from this work, access www.copyright.com or contact the Copyright Clearance Center, Inc. (CCC), 222 Rosewood Drive, Danvers, MA 01923, 978-750-8400. For works that are not available on CCC please contact mpkbookspermissions@tandf.co.uk

For Product Safety Concerns and Information please contact our EU representative GPSR@taylorandfrancis.com. Taylor & Francis Verlag GmbH, Kaufingerstraße 24, 80331 München, Germany.

Trademark notice: Product or corporate names may be trademarks or registered trademarks and are used only for identification and explanation without intent to infringe.

ISBN: 978-1-041-16390-9 (hbk)
ISBN: 978-1-041-16614-6 (pbk)
ISBN: 978-1-003-68544-9 (ebk)

DOI: 10.1201/9781003685449

Typeset in Times
by Newgen Publishing UK

Contents

Preface

This book centres on the decision-making processes of engineering and IT professionals, particularly as these processes emerge within heterogeneous environments such as human–computer interaction. Gaining a comprehensive understanding of the entire decision-making journey has become essential from multiple perspectives:

- Technological advancements engage data-driven society where decision-making is used to be a critical stage for innovative service delivery in industrial engineering, engineering management, smart manufacturing, and information systems engineering.
- The complex adaptive systems induce challenges through operational and situational uncertainties in decision-making.
- The decision-making process for human and machine intelligence would be crucial in human–computer interaction.
- Explaining a process flow of decision-making for virtual and remote uncertainties is demanded.
- Data- and model-driven approaches for rational decision-making in heterogeneous environments, particularly in engineering practices is critical and explanation of decision-making methods through use case scenarios is less addressed.
- The next generation decision-making through leadership, innovative and strategic framework is required to explain in the information age.

The journey mapping of decision-making in engineering practices is an extending demand for technology-driven society. To fulfil this requirement, academic institutions have subjects and programs that emphasize explaining decision-making. However, further explanations on decision-making with use case scenarios are critical for the following emergent grounds:

Firstly, the fundamental principles and ethical values of decision-making are evolving alongside technological advancements, which can be understood through the lenses of cognitive and emotional engagement. Few existing texts comprehensively address the decision-making process while integrating human factors, situational awareness, and critical thinking within structured decision-making frameworks. The ongoing paradigm shift in user–platform interaction capabilities necessitates further exploration and instructional guidance to help engineers and IT professionals develop the understanding and managerial skills required to make rational, judgment-based decisions and effectively navigate uncertainty.

Secondly, advanced design functionalities and system requirements are frequently changing the nature of interactions that engages a new direction in explaining decision-making through critical cognitive biases, human factors, machine behaviour, and heuristics. For example, digital touchpoints in retail platforms or artificial intelligence-based recommendations can influence decision-making through cognitive biases. In another instance, different heuristic tools like Figma or WordPress can influence designers' decision-making through advanced recommended features.

Digital platforms are becoming more popular to deliver services or products that creates uncertainties through virtualization, predictive analytics through artificial intelligence, and the need for touch can impact decision-making. This dynamic has come on the development of cognitive and emotional involvements – which had not been fully recognized by the information systems (IS), engineering management, industrial engineering, and business process management in the human–computer interaction paradigm.

Fourthly, engineering and IT professional practices emerge through advanced technologies, and it creates a converged field of study that accommodates multidisciplinary engagements. The new scenario-based use cases are complex in nature that demand further explanation with analytical tools like decision tree, influence diagram, and stochastic modelling. To deliberately satisfy the judgmental decision-making of engineers and IT professionals, universities and organizations can accommodate all the relevant theories and models like Bayesian networks, Markov decision process, multi criteria decision-making through AHP and TOPSIS, and group decision-making through Delphi method in the decision-making process. These analytical applications will support engineering professionals to develop their rational decision-making in factory operations, vendor selection, project implementation, supply chain management, and design thinking processes.

Finally, emerging technologies like artificial intelligence, virtual/augmented reality, internet of things, big data analytics, cloud computing, 5G, etc. are abruptly changing users' preferences, and it engages human and machine intelligence. Academics, professionals, researchers, industry leaders, governments, and regulators are facing challenges to cope up with the fastest growing trend in this digital transformation, where leadership, strategic, and innovative decision-making through Cynefin framework Roger's model, Double Loop Learning, etc. can explain the next generation decision-making process in engineering practices.

The objective of this book is to deliver guiding principles for decision-making for current digital landscape to engineering and IT students, academics, and professionals, accommodating it through a journey mapping of decision-making that reflects a complete guideline for IS, engineering management, industrial engineering, IT, and business analytics departments. A key consideration of the book is about real-world scenarios, examples, use cases, and analytical frameworks that can support students, engineering and IT professionals, practitioners, designers, and developers in understanding the whole journey mapping of decision-making, particularly in complex and uncertain situations.

At the postgraduate level or professional engagement, this book is well suited to references on information system analysis and design, cognitive systems, user experience research and design, engineering management, business analytics, and applied research. In addition, a short course or tailored course related to UX, digital marketing, cognitive processing, and engineering practices can adopt this book as a course reference. In both cases, this book will help students, academics, and professionals to enrich their theoretical, conceptual, managerial, and practical capabilities for complex adaptive information systems.

To conclude, it is my analogy that journey mapping of decision-making plays a critical role for engineers and IT professionals in the human-computer interaction paradigm that makes a different dynamic in information systems engineering aimed at several complex adaptive systems, not only autonomous driving, intelligent transportation systems, smart logistics, digital therapeutics, and robotic arm applications, but also virtual agents in interactive gaming, tourism, retail, and smart manufacturing.

Zian Shah Kabir
Sydney, New South Wales, Australia

Author

Zian Shah Kabir, PhD, is an experienced information system specialist with over 17 years of experience in the telecommunications industry. Currently, Dr. Zian is a research analyst and lecturer at the University of Technology Sydney, Australia. He is also a sessional lecturer at Macquarie University, the Lincoln Institute of Higher Education, and the Victoria Institute of Technology.

Dr. Zian earned a PhD in information systems engineering at the University of Technology Sydney, a master's in information and telecommunication engineering at the Korea Advanced Institute of Science and Technology (KAIST), and a bachelor's in electrical and electronic engineering at the Chittagong University of Engineering and Technology (CUET). He complemented his formal education with postgraduate diplomas and professional training in information technology.

After graduating, Dr. Zian joined the Bangladesh Telecommunication Regulatory Commission, serving as a telecom expert in various positions. His major responsibilities at the Telecom Regulator included governing the policy frameworks and formulating national standards to build a resilient and inclusive digital society. Dr. Zian has contributed as an international expert in international organizations (a focal point of the International Telecommunication Union). He has gathered vast experience in the communication industry and worked in several working groups to formulate standardization and recommendations. Dr. Zian has attended several workshops and seminars as a session speaker.

Dr. Zian's research spans artificial intelligence, user experience, digital transformation, virtual interactions, decision-making, and emerging regulatory perspectives. Throughout his career, he has led and contributed to numerous national and international working groups, delivered talks at workshops, conferences, and seminars, and developed strong expertise in quality assurance, system design, UX research, mixed-method research, cognitive studies, and large-scale information systems projects. His professional affiliations include the Institution of Engineers, Bangladesh (IEB) (fellow), KAIST Global Commercialization Center (GCC), the International Council on Systems Engineering (INCOSE), the Systems Engineering Society of Australia (SESA), the Australian Computer Society (ACS), the Commonwealth Engineers Council, and the Association for Information Systems (AIS).

Acknowledgements

The development and successful accomplishment of *Optimizing Decision-Making in Engineering: A Journey Mapping Perspective* has been made possible through tremendous efforts, valuable insights, and dedication of many persons and organizations. In the course of authoring the book, I learned more about decision-making, human factors, cognitive biases, situational awareness, and applications following theories and models. I gained practical knowledge from several lectures on decision-making, professional practices, IT professionals and society, and technology research preparation, which I had the privilege of delivering it in different academic and professional institutions.

I would like to express my thanks to those colleagues both from my academic and professional affiliations who supported and encouraged me during the whole journey. Many other professionals and academics in my personal connections have given their valuable times to review and endorsed their comments to revisit the book. Their review comments were invaluable in the journey process.

I would like to express my gratitude to all authors whose scholarly writings inspired me to deliver valuable works, particularly in decision-making form the intellectual foundation of knowledge. Their research outcomes, future direction, and contribution to developing knowledge in the emerging context of decision-making in the real-world applications that have made this book as a deep and intense resource with interdisciplinary fields.

Special thanks are conveyed to different organizations, governments, and regulators those who published their industry reports as open data, which supported me a lot to relate with decision-making situations and extended my use case analysis to make it more useful and resourceful for academics and professionals as their ready references. I am also thankful to industry leaders who contributed through their valuable suggestions and shared industry problems in the information age. Their professional rigor and research knowledge have deepened that the chapters mounted in this book enrich to the highest standards of scholarly excellence.

I am also truly glad to the professional communities and organizations that have collaborated in the research of cognitive and data-driven decision-making. Their valuable supports and collegial engagement have significantly enhanced the deliberations in this work.

A special thanks is conveyed to the technical experts, development editors, and book editors at the Taylor & Francis for their professionalism, guidance, and supports during every stage of publications. Their commitment and dedication have been influential in transforming this collaborative effort into an organized, systematized, and impactful publication.

Finally, I like to acknowledge the unwavering and valuable support of colleagues, mentors, and family members who have inspired me to contribute for the successful development of this project. Their backing and perceived social norms influenced me in excelling the authorship for *Optimizing Decision-Making in Engineering: A Journey Mapping Perspective*.

I hope that this book will assist to information systems community as enormous values to the continuing discourse on cognition, human factors, heuristics, situational awareness, theories and models, and emergence of technologies in engineering decision-making practices, inspiring further research collaboration and practical applications in this fast-growing field of study.

1 Fundamental Principles in Decision-Making

1.1 INTRODUCTION

In the contemporary era, problem-solving is often challenged by significant uncertainty, making rational decision-making (DM) increasingly complex. Situations such as designing autonomous vehicles, selecting an appropriate vendor for information system implementation, or choosing the correct operational strategy for an oil refinery represent high-stakes contexts that require careful judgement. When developing or managing complex adaptive systems, it is essential to recognize and address the full range of uncertainties that may influence outcomes. Effective DM in such environments draws upon a set of guiding principles integrated with disciplines including information systems engineering, risk and safety management, cybersecurity, cognitive science, and project management.

This chapter initially explains ethical values, mindfulness, and morale that act as enabling factors for engineers and IT professionals in making valuable decisions. The rational DM is a cognitive state that guides professionals to choose a best option from a set of alternatives based on different situations. For instance, a complex adaptive system engages uncertainties, and small deviations in DM can create unwanted consequences in operational activities. Therefore, professionals must follow ethical and process-oriented DM principles to avoid adverse consequences to the society.

In the information age, human–computer interactions within complex adaptive systems are shaped by uncertainty, with a range of theories and models explaining how these interactions influence DM. This chapter emphasizes the ethical dimensions of human DM by examining the roles of cognitive biases, human factors, machine intelligence, hallucinations and algorithmic bias, alongside core ethical values. It identifies key sources of bias, including representativeness heuristics, overconfidence, cognitive tunnelling, primacy and recency effects, framing and priming biases, and the illusion of control – that can significantly distort human judgement. Furthermore, the chapter demonstrates how DM principles and structured tools can be used to interpret and synthesize information intuitively, thereby enhancing situational awareness and supporting more informed decisions.

This chapter explains fundamental principles of DM through ethical dilemmas, mindfulness, morale values, and strategical approaches. The international council on systems engineering (INCOSE) developed a rule of practices for systems engineers

DOI: 10.1201/9781003685449-1

that can support them to follow standard procedures in professional practices. Further, different ethical approaches like Goal-Duty-Right-based analysis, 13 questions considering ethical actions, and four-step approach are fundamental principles in DM. Moreover, institutions of engineers and computer societies like Engineers Australia and Australian Computer Society (ACS) have developed and issued code of professional conduct that can enhance rational DM in professional practices.

1.2 PRINCIPLES OF DECISION-MAKING

1.2.1 Ethics, Mindfulness, and Morale

Ethics is a set of rules acknowledged in a respective class (society, religion, business, professional group, community, etc.) to perform their duties and actions maintaining a unique characteristic of a specific group and culture. The class or group of people issue an ethical code of conduct to obligate their members to do right and ethical acts for the betterment of the society. For instance, Engineers Australia, INCOSE, and ACS have issued ethical code of conduct for professional members to maintain their integrity and professionalism.

Ethics are critical in engineering practices as these codes of practices make professionals obligated for their course of actions and influence them towards rational DM. The ethical code of conduct maintains professional integrity, honesty, and dignity in DM that can support professionals to tackle critical situations while conducting their activities. However, it is imperative to make the DM more sustainable, considering a long-standing effect to the society. Therefore, a comprehensive set of DM principles can deliver judgemental, sustainable, measurable, achievable, ethical, relevant, innovative, and time-bound decisions in executing complex adaptive systems.

Then, mindfulness is a critical state that builds awareness to avoid conflict of interests and uncertainties. Unconsciousness and a lack of attention can lead to unethical behaviour through self-deception, biases, and self-serving cognition. Therefore, a lack of mindfulness can act as a catalyst for perplexing DM, which can deliver unethical behaviour in engineering practices.

In another context, people inherit moral values from their society, family, and culture that develop their ideology to conduct right or wrong activities. Morals are habitual perceptions or strong beliefs to perform right or wrong actions (Kant, 1785). They completely relate to an individual's ideology and norms. The moral reasoning engages awareness, intent, behaviour, judgement, and biases that create moral obligation to perform activities in a critical situation. Therefore, DM in engineering practices is a critical process that contains uncertainties, risks, complexities, biases, ethics, and moral values.

1.2.2 Semantic Theoretical Background

The ethical DM theory presents a model to explain the ethical DM as a process of delivering rational decision through moral reasoning (Schwartz et al., 2016). The ethical lens on DM theories support in understanding values, fair, transparent, and moral perspectives for professional practices.

1.2.2.1 Utilitarianism Theory

Bentham initially introduced the Utilitarianism Theory to explain the principles of morals and utilitarianism in the DM context (Bentham, 1789). This is a teleological or consequentialist theory, explaining that the moral action in choosing the right option depends on its valuable outcomes through consequences of actions. The DM process assesses the weighing scales of benefits over harms in every circumstance.

In DM perspective, utilitarianism pushes professionals to choose best option from available alternatives based on perceived experiences, social values, efficiency, and greater interest to deliver optimum performance. Although it does not depend on virtue or target, the utilitarian approach supports decision-makers to prioritize tasks that can provide an optimum level of goods.

1.2.2.2 Deontology Theory

Deontology, rooted in Immanuel Kant, claims that morality depends on loyalty to professional acts, duties, organization rules, and international best practices, not consequences of acts (Kant, 1785). Moral actions are judged by whether they follow rational, universal laws such as honesty, fairness, respect, and obligations to others.

A deontological or categorical approach underlines doing the "right thing" grounded in DM principles, even though the result is unprecedented. For example, deontological values for cellular mobile service providers are to protect subscribers' privacy and it is ethically wrong to share their personal identifiable information to telemarketers without getting prior consent.

Ethical decisions are grounded in duties such as truth-telling, respecting autonomy, and fulfilling commitments. From a moral perspective, this model values integrity and rule consistency. While not virtue-based, it aligns with virtuous behaviour by encouraging moral discipline and principled conduct.

1.2.2.3 Rights Theory

The rights-based ethical theory originates from moral perspective, political system, and social human rights. This theory explains that an individual inherits rights to maintain autonomy, privacy, dignity, and liberty. The good ethical conduct follows properly to protect all these rights. For instance, corrosive or manipulative advertisements through digital nudging in a mobile app can force customers to act in a certain way that interferes autonomy to take their rational decisions with reasoning.

In right-based ethical DM, it requires professionals to respect individual's choices regardless of coercive benefits. The theory uprights the values of respect, transparency, and judgemental process throughout DM. It is remarkably significant for professionals those who shall have to maintain data privacy, consent matter, or protect special need groups.

1.2.2.4 Virtue Ethics Theory

The great philosopher Aristotle explained virtue ethics, which aims to represent a character with moral values rather than leading systems, outcomes, or rules (Aristotle, 1894). The theory accentuates on virtues that combine a unique character with modesty, honesty, compassion, prudence, humanity, and strong courage that dominate an individual to take decisions rationally and ethically.

Virtue-based DM is critical for professionals in maintaining ethical values, which focus on developing moral character with a moderate balance between emotions and rationality. Usually, an individual develops virtuous ethics with moral wisdom through long-time engagements. It contributes to judgemental DM with ethical standards in professional practices.

1.2.2.5 Moral Intuitionism

In critical uncertainties, individuals apply moral intuitionism through instinctive judgements that do not need proper reasoning for taking immediate actions. Philosopher Moore and psychologist Haidt argue that moral decisions sometimes come from intuitive responses experienced by culture, existing knowledge, and perceived emotional engagements (Haidt, 2001; Moore, 1903).

In ethical DM, moral intuitionism plays a crucial role of delivering instant reactions through ethical values. The moral intuitionism can take a decision against accepting proposals that are considered to be rational in previous instances. Morally, intuition can support in delivering quick judgments, particularly in uncertainties or uneven situations. For example, cultural characteristics such as uncertainty avoidance or power distance can influence an individual's perception in adopting advanced information systems. However, moral intuition can also have an impact on biasing decisions in new use case scenarios. Therefore, the ethical DM through moral intuition can address a comprehensive approach to engage reflective knowledge, perceived experiences, evidence-based calculations, and moral reasoning.

1.2.2.6 Bounded Rationality

Bounded rationality, developed by Herbert Simon, challenges the classical assumption that humans make fully rational, optimized decisions (Simon, 1955). Instead, it argues that decision-makers operate under constraints such as limited information, cognitive capacity, time pressure, and organizational norms. Because these constraints "bound" our rationality, people often rely on heuristics, intuition, simplified models, or "satisficing" – choosing an option that is good enough, rather than perfect.

Bounded rationality reframes DM as a human, context-driven process rather than an idealized rational calculation. It emphasizes that:

- Ethical DM requires supportive structures to reduce bias.
- Moral judgment requires humility and sincere reasoning within limits.
- Virtuous character strengthens judgment where rules and data are insufficient.

Thus, bounded rationality complements other ethical theories by showing how decisions are made in practice, and how ethical awareness and virtues can improve the moral quality of those decisions despite inherent cognitive limits.

1.2.3 Decision-Making Theories

The DM process begins through interactions between environment and agent. Agents take observations by interacting with the environment, and agents perceive experiences in performing actions. However, DM is not just a course of action, but also it deals with uncertainties and explains the outcomes of information processing inside the brain. The classical DM theory like the Expected Utility Theory (EUT) followed normative model to choose the highest expected utility through rational thinking (Neumann & Morgenstern, 1955). The normative model considers a set of actions, a set of available events, a set of consequences realized for each sequence of action and event, and a set of probabilities for each combination. The EUT utilized mathematical formulas to identify the best feasible option in DM. However, Tversky and Kahneman introduced framing bias as a cognitive bias and then explored the relevance of descriptive models through cognitive biases and heuristics with risk aversion to explain the judgemental DM under uncertainty beyond economic values (Kahneman & Tversky, 1979).

With the advent of technology, researchers extend the philosophical viewpoints of DM by incorporating emerging constructs that would be more applicable for real-world industrial applications. This chapter reiterates the theoretical grounds for identifying DM processes in relevance to human-computer interaction (HCI) and Information Systems (IS) execution. Modern DM theories underline that decisions in organizations do not depend fully on rational models; rather, they address cognitive constraints, peer influences, and contextual restraints. The modern society follows an integrated DM framework that associates behavioural DM theories (such as utilitarianism theory and prospect theory) with technology adoption and evaluation-based DM models (such as Technology Acceptance Model, Unified Theory of Acceptance and Use of Technology, Task–Technology Fit, Expectation-Confirmation Model, and Stimulus-Organism-Response framework). Together, these advanced theories provide comprehensive frameworks to understand DM through design choices, user adoption, user experience, behavioural response, and retention mechanism in real-world scenarios.

In the information age, providers are trying to involve customers' preferences with the probability analysis of loss and gain in achieving goals. The prospect theory explains that higher level of gains leads to a higher probability of risk aversion, whereas a lower level of loss invites risk-taking situation (Kahneman & Tversky, 1979).

BOX 1.1 AN EXAMPLE SCENARIO WITH THE PROSPECT THEORY IN DM

We can explain the Prospect Theory for an example with two software programs in a metro rail project, where program A has guarantee to save 600 passengers and program B has a 30% chance of saving all 1000 passengers, and a 70% chance that nobody would be saved in the gain frame.

In the loss frame, program A can result in losing 400 passengers' lives and program B has a 30% chance that no one will die, and 70% chance for losing all lives (1000 passengers).

In the eye of DM perspective, positive framing with saving 600 passengers can claim for risk averse (concave of gains) and negative framing with the possibility of sacrificing 600 lives can push towards risk-seeking situation (convex of losses).

Therefore, professionals can consider both positive and negative framing with associated risks rather than only numerical values before taking critical decisions for project development or information system implementation.

Decision-makers attempt to apply DM frameworks considering real-world scenario contexts. Technology Acceptance Model (TAM) and Unified Theory of Acceptance and Use of Technology (UTAUT) are popular frameworks that can deliver valuable insights and key performance constructs such as perceived ease of use, perceived usefulness, peer influence, performance expectancy, effort expectancy, and facilitating condition and enabling professionals to understand user adoption in software adoption, HCI usability assessment, or other information systems implementation projects. Professionals can apply those DM models in executing user acceptance test (UAT), performance monitoring, user experience (UX) design, Low and Hi fidelity Prototype, and iteration processes in product development life cycle.

In the context of technology utilization, Task–Technology Fit (TTF) describes that users accept an information system while it associates with their workflows, task requirements, and relates to enhance performance. The technology adoption relates on how it fits between user requirements to accomplish tasks and functionalities of analytical tools. The theory highlights that technological functionalities are not sufficient in the adoption process, where socio-cultural integration and contextual considerations are crucial for successful IS implementation. Professionals can apply an iterative tailored approach to evaluate peer influence and technology readiness to move forward in information systems implementation.

The Expectation–Confirmation Theory (ECT) suggests how individuals perceive confirmation as a form of satisfaction by mitigating their expectation from products, services, or information systems. The level of satisfaction is assessed through perceived performance of products or services in comparison with initial expectations from user end. The ECT explains four sequential constructs in describing the DM process through expectations, perceived performance, confirmation/disconfirmation, satisfaction, and continuance intention. In information system research, industries and academics can apply ECT in various contexts to explain post-adoption user satisfaction, behavioural intentions, and system improvement assessment, particularly in the DM phases that relate to continuing to use an information system.

The Stimulus–Organism–Response (S-O-R) framework was initially introduced as a psychological model that describes how environmental stimuli factors impact individuals' cognitive and emotional states; consequently, the perceived organisms influence behavioural intentions. The S-O-R has enhanced a theoretical lens in extending research to explore DM through technology adoption, cognitive acceptance or dissonance, continuance intention, and behavioural responses in IS implementations. S-O-R is especially effective for identifying critical stimuli factors and how system, user interface, or virtual interactions effect user cognitive engagements, emotions, attitude, trust, satisfaction, and behavioural intentions to adopt or continue using a technological system. DM theories and models with outcomes and descriptions are described as shown in Table 1.1.

TABLE 1.1
Decision-Making (DM) Theories and Models

Theory/Model	Authors	Constructs	DM Outcomes	Description
Technology Acceptance Model (TAM)	Davis, F. D. (1989)	Perceived usefulness, perceived ease of use, attitude	Behavioural response, actual system use	To explain how users come to accept and adopt a technology
Unified Theory of Acceptance and Use of Technology (UTAUT)	Venkatesh et al. (2003)	Performance expectancy, effort expectancy, social influence, facilitating conditions	Behaviour intention	To provide a highly predictive model of explaining technology acceptance towards actual usage behaviour
Task–Technology Fit (TTF)	Goodhue and Thompson (1995)	Task characteristics, technology characteristics, task technology fit	Adoption intention, continued intention	To explain how well a technology supports the tasks users must perform by considering functionalities, user requirements, and benefit from a system
Expectation–Confirmation Theory (ECT)	Oliver (1980)	Expectation, perceived performance, confirmation, satisfaction	Continuance intention	To explain how individuals perceive satisfaction through confirmation of initial expectations and perceived performance of a product or system
Stimulus–Organism–Response (S-O-R) framework	Mehrabian and Russell (1974)	Stimulus, cognitive and emotional states	Continuance intention, purchase intention, word of mouth	To explain decision-making through stimuli, cognitive, emotional, and behavioural responses through interactions, attitude, trust, satisfaction, and continued use in digital environments

Professionals can apply DM theories and models to design information systems by accommodating perceived usefulness and ease of use, cognitive limits, functionalities, expectations, and behavioural responses that support to deliver strategic recommendation in adopting technological systems.

BOX 1.2 AN EXAMPLE SCENARIO WITH THE S-O-R FRAMEWORK

Academic researchers or industry professionals can apply the S-O-R framework to explore DM results and valuable insights in designing digital platforms. Professionals can conduct focus group interviews to explore new stimuli factors through qualitative analysis (using NVivo) in an augmented reality (AR) environment. Then they can develop a S-O-R model explaining cognitive and emotional constructs, particularly attitude, trust, and satisfaction that can be linked to continuance intention. The study can collect survey data from online users and extend a quantitative analysis (using SPSS or Smart PLS) to establish relationships among constructs. For example, considering an AR mobile platform, professionals can execute partial least square – structural equation modelling (PLS-SEM) satisfaction has more influence than attitude and trust towards continuance intention through hypothesis testing.

1.3 PROCESS FLOW OF DECISION-MAKING

Figure 1.1 demonstrates an integrated DM process flow that connects fundamental principles, semantic and advanced theoretical frameworks, contextual factors, and evidence-driven outcomes in taking ethical decisions. The top section (DM principles) appears how DM begins with broad principles such as ethics, mindfulness, and professional codes of conduct that guide the professional's mindset. These principles are then connected to DM theories and models, denoted by nodes like Semantic Theory and Modern DM Theory. The process flow visually shows how these theories are integrated to form a rational DM foundation. If expected, the process can loop back through reiteration process to earlier theoretical steps, establishing that DM is iterative rather than linear.

The middle and lower sections represent how theoretical integration moves into real-world application considering contextual use-case scenarios. Contextual application evaluates socio-cultural and organizational impacts, pursued by an approval check to confirm the chosen method that fits the context. If applicable, the decision advances into evidence-driven DM, where evidence-based outcomes, moral reasoning, and cross-validation with existing knowledge are considered. This leads professionals to taking ethical decisions that support them to execute the tasks judgementally. Overall, the figure represents a comprehensive and instinctive DM process flow where DM principles, theories, context, and evidence-based outcomes organize in association with well-justified decisions.

Professionals can apply the process flow map in different contexts like communication, healthcare, retail, agriculture, logistics, etc. However, they can contextualize

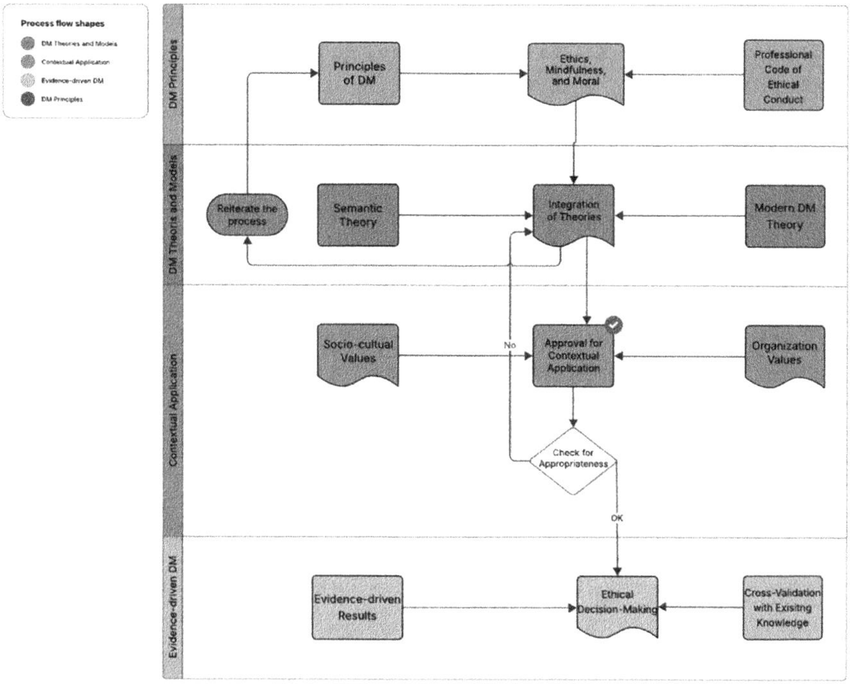

FIGURE 1.1 Process Flow Map for Decision-Making (DM).

relevant DM theories and models to justify the applicability of technology adoption and DM perspective. The critical phase is to integrate socio-cultural values and organization/industrial code of practices to reach into ethical DM.

The ethical DM process is a critical step for an organization, particularly in executing information systems considering value chain and constraints in various contexts. Even though the process is time-consuming, it is essential to follow a standardized process to avoid design inappropriateness, reputational damages, risks, and additional costs for re-establishment.

Professionals can follow the standardized process flow map during the whole information system implementation and co-create values through ethical DM. The semantic theoretical background, modern theoretical models, and advanced analytical approaches can play a pivotal role in achieving the project goals through successful deliberations.

BOX 1.3 EXAMPLE CASE SCENARIO ON DM USING PROCESS FLOW MAP

For example, a mobile service provider is going to implement an agentic artificial intelligence (AI)-centric customer support system to reduce wait times, cut operational costs, and enhance customers' perceptions. Using the DM framework, the process flow map starts with fundamental DM principles. Professionals can first follow principles of professional conduct such as transparency, obligation, fairness, accountability, and commitment to ensure that the AI will not delude customers or violate data privacy. The initial planning considers ethics and mindfulness, confirming that the system complies with the telecom regulatory standards and customer protection-related regulations. Then, these regulatory and ethical principles follow a theoretical evaluation process by reviewing Semantic Theory and Modern DM Theory, which can help decision-makers to interpret valuable information, moral reasoning, and integrate multiple theories (e.g., utilitarian theory, rights theory, stakeholder theory). If discrepancies occur, the process can loop back to reiterate the assessment.

In the next course of action, the company can proceed to contextual application, assessing socio-cultural values (e.g., customers' preferences, cultural characteristics, social norms) and organizational/industrial values (e.g., customer retention, rules for mobile/customer association). The anticipated AI system is assessed in terms of cultural orientation, accessibility, and correlation with brand reputation. An appropriateness check can warrant that the AI does not make inequities in respect of gender, race, and accessibility issues. For example, excluding customers' orientation with digital gadgets and functionalities of applications. Once the system is executed, the company can go through the Evidence-Driven DM phase, conducting experiments, focus-group interviews, and A/B testing, analyzing key performance indicators, and cross-validating outcomes with existing knowledge from previously available data. The evidence-driven results can suggest on the reductions in call-centre load and maintain to improve the grade of service consistency, leading to a final ethical DM to implement the system.

1.4 SUMMARY AND DISCUSSION

Ethical DM can be realized from different perspectives, depending on the user requirements and available solutions. This chapter emphasized on applying ethics, mindfulness, and moral in the DM process. Further, semantic theories and models are critical to understand the principles of morals and utilitarianism, loyalty to professional duties and organization rules, moral perspective, political system, social human rights, virtue, instinctive judgements through moral intuitionism, and bounded rationalities.

More specifically, the chapter includes advanced DM theories and models like expected utility theory (EUT), prospect theory, technology acceptance model (TAM), unified theory of acceptance and use of technology (UTAUT), task–technology fit (TTF), ECT, and S-O-R framework to explain the ethical DM process through multiple constructs that can allow professionals to understand the valuable insights in executing and implementing information systems.

Finally, a process flow map and a few example use-case scenarios can guide professionals, governments, regulators, solution providers, standardized institutions, organizations, and technology vendors in making ethical and moral decisions considering codes of conduct, theoretical grounds, relevant frameworks, process flow maps, and thorough evaluation techniques. In the next chapter (Chapter 2), we begin the detailed focus on explaining a journey map of DM in engineering practices to determine a way forward for professional development.

BIBLIOGRAPHY

Aristotle. (1894). *Ethica Nicomachea.* Edited by I. Bywater. New York: Oxford University Press.

Bentham, J. (1789). *An introduction to the principles of morals and legislation.* T. Payne.

Davis, Fred D. (1989). Perceived usefulness, perceived ease of use, and user acceptance of information technology. *MIS Quarterly*, 13(3), 319–340. https://doi.org/10.2307/249008

Goodhue, Dale L. & Thompson, Ronald L. (1995). Task-technology fit and individual performance. *MIS Quarterly*, 19(2), 213–236. https://doi.org/10.2307/249689

Haidt, J. (2001). The emotional dog and its rational tail: A social intuitionist approach to moral judgment. *Psychological Review*, 108(4), 814–834. https://doi.org/10.1037/0033-295X.108.4.814

Kahneman, D., & Tversky, A. (1979). Prospect theory: An analysis of decision under risk. *Econometrica: Journal of the Econometric Society*, 47(2), 263–291. https://doi.org/10.2307/1914185.

Kant, I. (1785). *Groundwork of the metaphysics of morals.* Cambridge University Press (translated editions vary).

Kim, B. J. H. (2024). Aristotle's nicomachean ethics on virtue competition. *British Journal for the History of Philosophy*, 32(1), 1–21. https://doi.org/10.1080/09608788.2023.2243605.

Mehrabian, Albert & Russell, James A. (1974). *An Approach to Environmental Psychology.* Cambridge, MA: MIT Press.

Mill, J. S. (1863). *Utilitarianism.* Parker, Son, and Bourn.

Moore, G. E. (1903). *Principia ethica.* Cambridge University Press.

Oliver, Richard L. (1980). A cognitive model of the antecedents and consequences of satisfaction decisions. *Journal of Marketing Research*, 17(4), 460–469. https://doi.org/10.2307/3150499

Ross, W. D. (1930). *The right and the good.* Oxford University Press.

Schwartz, Mark S. (2016). Ethical decision-making theory: An integrated approach. *Journal of Business Ethics*, 139(4), 755–776. https://doi.org/10.1007/s10551-015-2886-8

Simon, H. A. (1955). A behavioral model of rational choice. *The Quarterly Journal of Economics*, 69(1), 99–118. https://doi.org/10.2307/1884852.

Simon, H. A. (1957). *Models of man: Social and rational; mathematical essays on rational human behavior in a social setting.* Wiley.

Venkatesh, Viswanath, Morris, Michael G., Davis, Gordon B. & Davis, Fred D. (2003). User acceptance of information technology: Toward a unified view. *MIS Quarterly*, 27(3), 425–478. https://doi.org/10.2307/30036540

Von Neumann, J., & Morgenstern, O. (1955). *Theory of games and economic behaviour* (6th ed.). Princeton University Press.

2 A Journey Map of Decision-Making in Engineering Practices

2.1 INTRODUCTION

Decision-making is a foundational function within any organization, particularly when choices carry substantial risk, uncertainty, or the potential for loss of human life. Organizational decisions are closely tied to stakeholder trust, long-term performance, and ethical conduct, especially in complex and high-pressure contexts such as critical engineering operations. In these environments, decision-making not only determines mission success but also follows a structured journey through processes designed to safeguard professionals and preserve organizational integrity.

Judgemental and rational decision-making are therefore central to the safety and sustainability of engineering systems. The quality of decisions made by developers, engineers, designers, operators, and regulators has a direct and profound impact on public welfare. In sectors characterized by high-risk activities, the ability to exercise sound judgement and commit to well-informed decisions plays a decisive role in shaping outcomes amid complexity and uncertainty.

Making important decisions and carrying them out are crucial to determining alternatives in complex and unpredictable situations, particularly in engineering practices with elevated levels of risk. Making engineering decisions is commonly characterized by ambiguity, time bias, organizational requirements, and goals. In this case, decision-makers are expected to process technical information, strike a balance between organizational and social demands, and predict safety results without breaking cognitive and systemic constraints. A journey map of decision-making in engineering practices engages cognitive, model, and data-driven models that can allow professionals to make rational and judgemental decisions.

Rational decisions are essential to organizations because it shapes long-term performance, ethical standards, and strategies. It makes a journey process that may involve cognitive bias or logical contexts in the decision-making. Instead, they are impacted by heuristics and cognitive biases, which are mental shortcuts that can be simplified by accommodating cognitive, model-, and data-driven models. Heuristics can be helpful in situations where there is uncertainty. However, if they are not used properly, they can lead to incorrect logic, unethical blind spots, and suboptimal results. Therefore, promoting thoughtful judgment and moral decision-making in modern businesses follows established models in recognizing and reducing these biases.

DOI: 10.1201/9781003685449-2

Decision-making represents the cornerstone of engineering judgment, as every stage of an engineering project can execute cognitive, model-, and data-driven models, from conceptual design to operation, depending on how effectively professionals evaluate alternatives under uncertainty. These modular analyzes avoid poorly structured or biased decisions that can result in catastrophic consequences.

2.2 DECISION-MAKING MODELS

2.2.1 Applications of Models in Decision Context

This chapter provides a detailed rationale for the selection and application of the three decision-making models – cognition-driven, model-driven, and data-driven that can be used to analyze decision-making situations. It also discusses each model's assumptions, inputs, applications, and unique contributions to understanding critical decision situations.

To show how theoretical frameworks provide insight into the fundamental causes of failure and offer suggestions for how organizations can stay clear of similar problems, this chapter critically connects to explain a journey mapping towards decision-making through three independent decision-making models. Professionals can follow these three models to analyze decision-making situations and can identify biases, heuristics, alternatives, and choose a best option to execute their tasks. By doing this, the chapter highlights the importance of combining structural, psychological, and cognitive approaches to decision-making in complex engineering environments.

The decision-making spectrum can be understood as a continuum shaped by three dominant approaches: cognition-driven models, model-driven models, and data-driven models, each representing a different balance between human judgment, analytical structure, and empirical evidence. In practice, effective organizational decision-making unfolds as a deliberate process that begins with a critical review of organizational activities and a careful analysis of stakeholders, followed by the identification of specific decision contexts that demand attention. Within these contexts, cognition-driven models draw on experience, intuition, and mental frameworks to interpret complex or ambiguous situations; model-driven approaches introduce structured analytical tools and formal frameworks to compare options systematically; and data-driven models rely on quantitative evidence and patterns extracted from data to guide choices.

Underpinning these analytical approaches is the rational decision-making model, which assumes that decision-makers or professionals are able to clearly define the problem, gather all relevant information, identify feasible alternatives, evaluate those alternatives objectively and ethically, and ultimately select the option that is both the safest and most effective. While this rational ideal provides a valuable foundation, the spectrum acknowledges that real-world decisions often require an integration of human cognition, analytical models, and data-based insights to navigate uncertainty and complexity.

Cognition-driven models emphasize the psychological biases that influence human judgement, while model-based approaches apply established decision frameworks to systematically organize and evaluate alternatives. This perspective reveals how heuristic reasoning and executive overconfidence can lead to the underestimation of logistical risks. By uncovering such psychological blind spots, cognition-based models support the adoption of strategies such as pre-mortem analysis and diverse, cross-functional decision-making. In contrast, model-driven approaches demonstrate how structured reasoning, formal evaluation, and risk framing can guide organizational communication and improve outcomes, validating the use of rational frameworks and adaptive cycles in managing reputational and operational crises.

Data-driven approaches, by comparison, offer a proactive means of identifying, quantifying, and preventing failures through evidence-based insights. By leveraging analytical methods and real-time data, these models enable organizations to anticipate risks rather than merely react to them. The integration of continuous learning loops further enhances decision intelligence, allowing organizations to refine their decision-making capabilities over time and strengthen resilience in complex and uncertain environments.

2.2.2 Visual Summary of Decision-Making Models

Collectively, decision-making models demonstrate how structured frameworks can enhance reliability, transparency, and accountability in complex and uncertain contexts, such as the implementation of large-scale infrastructure projects. The analysis also offers direction for future engineering practice, underscoring that sound judgement depends on the integration of rigorous technical analysis with ethical awareness and an ongoing reflection on human cognitive limitations. This integrated perspective is illustrated through a visual summary presented in Figure 2.1.

In the visual summary, it explains that a cognition-driven model supports professionals to extend situational analysis through psychological and behavioural factors that can shape situations, integrating theories such as bounded rationality, groupthink, overconfidence bias, and other relevant theories. A model-driven framework, on the other hand, can be used to assess and identify alternatives through crisis communication response, guided by rational decision-making principles and prospect

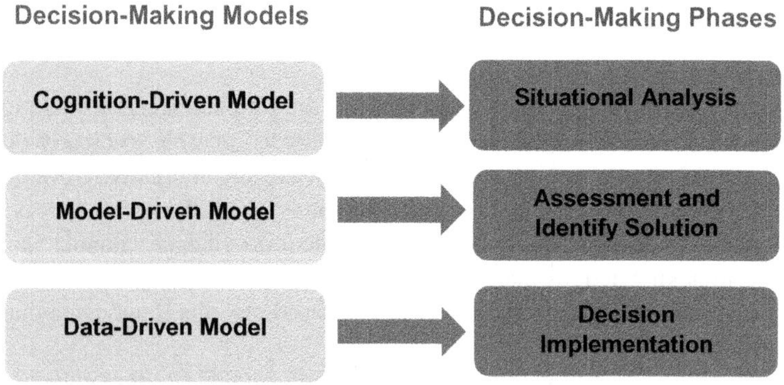

FIGURE 2.1 Visual Summary for Decision-Making Models and Phases.

theory. Finally, a data-driven lens can briefly explore to demonstrate how predictive analytics and real-time data can support evidence-based judgement through decision implementation.

2.3 COGNITION-DRIVEN DECISION-MAKING

2.3.1 Theoretical Foundation

Cognition-driven decision-making focuses on how people think, perceive, and process information when making choices. It recognizes that human decision-makers do not always act rationally; instead, they rely on mental shortcuts, emotions, experience, and social influences. This model draws from behavioural psychology and cognitive science to explain the biases and heuristics that can distort judgment. Factors such as overconfidence, groupthink, anchoring, authority bias, and normalization of deviance often lead to decisions that ignore critical evidence. The cognition-driven approach seeks to understand these mental processes and how they interact with organizational structures and cultures. By examining how individuals and groups form perceptions of risk, interpret data, and respond to pressure, cognition-driven models aim to identify weaknesses in decision systems that stem from human behaviour rather than technical deficiencies.

The cognition-driven decision-making explains psychological and cognitive processes through which individuals and organizations interpret information, assess alternatives, and select courses of action. Rather than assuming purely rational or quantitative behaviour, these models recognize that decision-makers operate under bounded rationality and frequently rely on heuristics. In high-risk industrial settings, such as chemical processing plants, cognitive limitations and biases strongly influence decisions, particularly under time pressure, uncertainty, and organizational stress. Viewing decisions through a cognitive lens helps explain why seemingly irrational actions such as deferring safety upgrades or ignoring early warning signals can occur, as these choices often emerge from constrained reasoning rather than deliberate negligence.

This perspective is underpinned by several key psychological theories. Herbert Simon's theory of bounded rationality explains that decision-makers tend to "satisfice," choosing solutions that appear acceptable within cognitive, informational, and organizational limits, even if they carry long-term risks (Simon, 1955). Kahneman's dual-process theory further suggests that under stress, individuals rely more heavily on fast, intuitive system thinking, while slower, analytical system reasoning is underused due to resource and organizational constraints. Prospect theory complements this by showing that decision-makers tend to overweight potential losses relative to gains, leading management to prioritize short-term financial survival over low-probability but high-consequence safety risks (Kahneman & Tversky, 1979). Together, these theories explain how economic pressure, cognitive bias, and organizational context can converge to produce risk-seeking behaviour that undermines safety in critical industrial operations.

Several psychological theories offer a structured lens for interpreting an organization's cognition-driven decision-making. Bounded rationality explains how managers, constrained by limited time, information, and cognitive capacity, opted to satisfice rather than optimize, accepting decisions that appeared sufficient without fully examining worst-case risks. Prospect theory further shows that an organization can frame decisions around avoiding financial losses rather than preventing environmental harm, encouraging risk-seeking behaviour such as reducing investment in safety. This tendency can be reinforced by the normalization of deviance, where repeated acceptance of departures from safety standards fostered complacency and made risky practices seem routine rather than exceptional. Finally, cognitive dissonance theory highlights how an organization's self-image as an environmentally responsible organization conflict with its actions, prompting legitimizing narratives aimed at reconciling corporate identity (Festinger, 1957).

The decisions analyzed in this chapter illustrate how different forms of reasoning operate in parallel, each requiring distinct yet complementary theoretical lenses. The cognition-driven model reveals how intuitive judgement, group dynamics, and confidence in established practices contributed to the absence of new pilot training or transparent system disclosure. In contrast, the model-driven framework highlights how external pressures – such as competitive forces, certification requirements, and resource constraints influenced strategic choices around system modification. Together, these perspectives demonstrate that organizational decisions rarely emerge from a single rational pathway; rather, they are shaped by the interplay between human cognition and structured, contextual constraints.

2.3.2 Cognition-Driven Models

Cognition-driven decision-making models are grounded in psychological theories that explain how individuals process information, interpret uncertainty, and act under pressure. Unlike purely rational models, it emphasizes the limitations of human cognition in complex, ambiguous, and high-stakes environments such as large-scale engineering projects. Decision outcomes are shaped by heuristic mental shortcuts that simplify problem-solving but also introduce systematic biases. Within engineering contexts, these include overconfidence bias, where experts overestimate their predictive accuracy, and confirmation bias, where they selectively address data that support prior assumptions. Anchoring and availability heuristics further distort judgment by focusing on initial estimates or memorable past incidents.

In cognition-driven decision-making model, it emphasizes on the use of the cognitive functions, such as perception, attention, memory, and reasoning to guide choices and evaluate complex situations. The model focuses on how individuals rely on intuition, heuristics, and mental shortcuts when faced with uncertainty, pressure, or incomplete information. Rather than assuming that decisions are always the result of rational analysis, this approach recognizes that judgement is often shaped by internal biases, organizational influences, and past experiences. In complex environments, cognitive factors can substitute for structured reasoning, particularly when time constraints or commercial pressures exist. This model is appropriate for examining decisions that emerge informally rather than through formal evaluation procedures.

The cognition-driven model integrates psychological theories to map the biases and heuristics influencing professionals' choices. It reveals how urgency bias, overconfidence, diffusion of responsibility, and other cognitive biases can dominate in making decisions under uncertainties. These cognitive limitations can lead to fragmented accountability and an underestimation of infection-control risks.

The cognition-driven decision-making model emphasizes the internal mental processes individuals rely on when making decisions, including perception, reasoning, intuition, memory, experience, and emotional responses. In practice, people often depend on these cognitive mechanisms rather than purely logical or data-driven approaches. This model is particularly relevant in situations involving time pressure, stress, limited information, and strong psychological influences. In high-risk domains such as medical emergencies or autonomous driving, cognition-driven decision-making helps explain how errors can occur even among highly experienced professionals.

Cognition-driven decision-making models examine how individuals and teams interpret, process, and respond to information under constraints of time, complexity, and uncertainty. According to the concept of bounded rationality, decision-makers rarely achieve perfect rationality because they operate with limited knowledge, imperfect information, and constrained time. Instead, they "satisfice" by selecting the first acceptable alternative rather than the optimal one.

The models of cognition-based decision-making investigate the role of psychological processes, heuristics, and biases in influencing decisions in an organization in a complex environment. In contrast to fully rational structures, they demonstrate how human constraints interrelate with organizational demands to lead to poor decisions.

In professional practice, cognition-driven models are applied to improve awareness of biases and promote reflective thinking. Tools such as structured decision reviews, cognitive bias training, and pre-mortem analysis are used to strengthen analytical reasoning and reduce the influence of intuition under stress. The goal is not to eliminate intuition but to balance it with analytical processes, ensuring that decisions are both informed and deliberate. This model is particularly valuable in contexts where uncertainty, time constraints, and emotional pressures are high, such as emergency response, aviation, or space exploration. It emphasizes that even highly skilled professionals can make poor decisions if psychological factors and social pressures are left unchecked. The cognition-driven model demonstrates that:

- Psychological factors can override engineering logic.
- Leadership overconfidence and framing can distort risk perception.
- Organizational echo chambers suppress corrective feedback.
- Absence of cognitive debiasing tools allowed unsafe assumptions to stand unchallenged.

2.3.3 Cognition-Driven Decision-Making Process

The cognition-driven process demonstrates that cognitive and organizational biases are central in making decisions. It reveals that expertise alone cannot guarantee rational

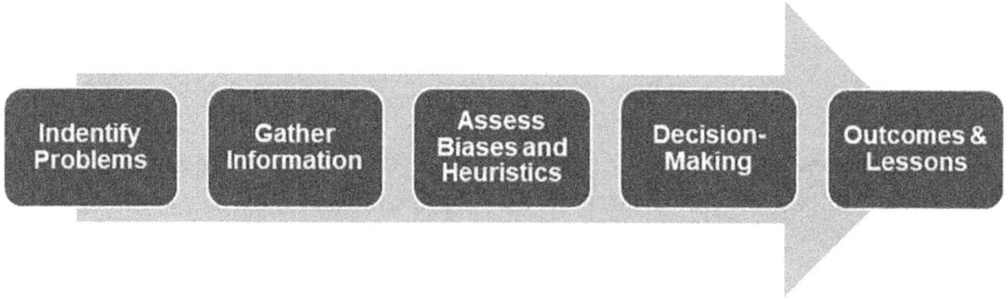

FIGURE 2.2 Cognition-Driven Decision-Making Process.

judgment when decision-making occurs within closed feedback environments and under institutional pressure. By examining decision situations through a cognition-driven lens, it becomes possible to explain why apparently irrational choices, such as cutting safety budgets or delaying critical warnings, may emerge under pressure. This chapter applies psychological theories and bias analysis to explain a cognitive decision-making process for critical situations.

The cognition-driven decision-making process as shown in Figure 2.2 emphasizes to identify problems by analyzing psychological and cognitive engagements through which individuals and organizations gather and interpret information. Then, the process includes a comprehensive assessment for biases and heuristics to evaluate alternatives and choose courses of action. The process moves beyond purely rational or quantitative perspectives in decision-making, recognizing that decision-makers often rely on heuristics and operate within the limits of bounded rationality. Finally, the process accommodates to evaluate outcomes through lesson learned. In high-risk industrial environments such as chemical plants or aeronautical industries, the cognition-driven process can identify cognitive constraints and biases that play a decisive role in shaping outcomes.

This cognition-driven process focuses on the decision-makers' cognitive characteristics and pinpoint the regions that lead to uncertainties. To show how mental shortcuts influence professionals' perception of the circumstances, the process incorporates psychological theories like the availability shortcuts and overconfidence bias. To investigate how decision-makers may improve their situational awareness and decision-making under pressure, the process can make professionals confident to follow the cognition-driven decision-making process through analytical instruments and strategies such as impact graphs, creative thinking, influence diagram, decision tree, and SWOT analysis.

Overconfidence and confirmation biases produce a false sense of control, while groupthink suppresses technical evidence dissenting. The analysis further indicates that the absence of structured reflection mechanisms such as scenario planning and team evaluations can allow biased reasoning to persist in irrational decision-making. By mapping these dynamics, the process highlights the importance of embedding cognitive resilience in engineering governance, where individuals are trained to recognize and challenge mental shortcuts. Debiasing strategies, including transparency in data sharing and routine cross-disciplinary audits can foster more mindful decision-making. The process underscores that preventing future infrastructure disasters requires integrating psychological awareness into technical systems, ensuring that engineering decisions remain both analytically sound and cognitively self-critical.

The cognition-driven choice-making approach looks at how people and organizations receive, process, and react to complex information, particularly under pressure and in uncertain situations. It focuses on human operators' psychology in the situation of stress, uncertainty, and limited rationality. The cognition-driven decision-making process reveals several critical judgement errors.

The cognition-driven process assumes that decision-makers operate under bounded cognitive capacity, limited data reliability, and institutional constraints. Inputs include technical inspection records, sensor readings, and visual assessments, but the interpretation of these inputs depends on experience and mental framing. The process presumes that organizational culture discourages questioning senior authority, thereby constraining corrective action. Another assumption is that uncertainty about material degradation was high due to the concealed nature of encased cables, leading engineers to rely on subjective judgment rather than quantitative evidence.

Limitations of the cognition-driven process include the difficulty of quantifying cognitive variables such as motivation, trust, and responsibility diffusion. Moreover, retrospective application cannot fully capture the emotional and political pressures that shaped real-time decision behaviour. Nevertheless, the model remains valuable as a diagnostic framework for identifying where decision support, bias-awareness training, or cognitive debiasing interventions could have altered the course of maintenance decisions.

2.3.4 Implications and Insights

The cognition-driven approach aims to explain why people and groups make irrational decisions because of their cognitive biases and heuristics that can invite uneven consequences. This chapter, which draws from cognitive psychology, focuses on dual process thinking, biases, heuristics, and ethical fading. The cognition-driven model supports professionals to understand how cognitive functions process information, make decisions, and rationalize conduct under uncertainties and confusion.

Cognitive decision-making frameworks focus on how people and groups see, understand, and behave in unpredictable, complex situations. Making decisions in the military and aviation industries frequently involves emotional stress, insufficient information, and extreme time pressure – all of which impair logical reasoning.

The cognitive model develops senses because it reveals the psychological processes that drive otherwise logical experts' unethical behaviour, including self-deception, confirmation bias, and escalation of commitment. It also emphasizes how perceptions of appropriate behaviour are shaped by the interaction between organizational culture and cognition.

Additionally, decisions are made within the framework of constrained morality, where moral considerations are accidentally disregarded because of framing effects or environmental signals, as acknowledged by cognition-driven analysis.

In the end, the cognition-driven model can be selected due to its ability to represent the human aspect of poor decision-making. It provides insight into how heuristics and biases affect professional's thinking, how moral considerations decrease as technical problems are solved, and how self-justification permits immoral behaviour to continue.

Cognitive models enable automatic, context-based behaviour in time-sensitive settings. But they break down when information is misunderstood or when biases take over. Cognitive models, which are based on psychological theory, perform best in situations that are unclear, time-sensitive, and have little access to data. They take advantage of professional intuition, life experience, and the ability to quickly identify well-known patterns. The crucial role that cognitive processing plays in making survival decisions in real time is demonstrated by the pilot's quick attempt to steer the crashing jet away from residential areas. However systematic bias can also be sustained by these same mechanisms.

For example, command-level leaders displayed both authority bias, which discouraged dissent from subordinates, and normalcy bias, which underestimated anomalies because of excessive exposure to routine success. These biases prevent reevaluating preconceived notions by limiting situational awareness. Additionally, limited rationality constrains commanders' ability to comprehend the entire range of risk information, which results in satisfying decisions rather than the best ones.

To put it briefly, cognitive models increase agility at the expense of consistency. Even highly skilled decision-makers can miscalculate probabilities in the absence of external data validation, especially when dealing with uncommon but disastrous events.

2.4 MODEL-DRIVEN DECISION-MAKING

2.4.1 Theoretical Foundation

Model-driven decision-making approaches rely on formalized frameworks, logical reasoning, algorithms, and quantitative models to evaluate alternatives and predict probable outcomes. This approach assumes that decisions should be grounded in data, structured analysis, and established theoretical models. Unlike cognition-driven models, which emphasize the psychological processes and inherent limitations of human reasoning, model-driven approaches attempt to overcome such limitations by systematically incorporating data, probabilities, and rational frameworks. These models enable decision-makers to quantify uncertainty, evaluate trade-offs, and anticipate long-term consequences.

Model-driven decision-making refers to approaches where choices are shaped by structured reasoning rather than intuition or habit. These approaches emphasize clarity of objectives, evaluation of alternatives, and the weighing of constraints before selecting a course of action. In organizational contexts, such decision-making often arises when long-term investments, competitive pressures, regulatory rules, and cost implications are involved. Unlike cognition-based reasoning, where unconscious bias and heuristics dominate, model-driven decisions tend to reflect trade-offs formed around organizational priorities and available resources.

The model-driven process often uses mathematical tools, decision trees, simulations, or probabilistic models such as Bayesian networks (BN). The BN model is useful for situations involving uncertainty and incomplete information. It works by combining prior knowledge or historical data with new evidence to calculate the probability of different outcomes. This allows decision-makers to continuously update their understanding of risk as latest information becomes available, leading to more objective and transparent decisions.

Model-driven decision-making aims to minimize human error by replacing intuition and subjective judgment with structured reasoning. By quantifying uncertainty and systematically analyzing relationships between variables, organizations can identify optimal courses of action based on evidence. For this complex planning scenario, a model-driven approach is appropriate to bring rigor and transparency to analyze.

Model-driven decision-making is about using established frameworks to help in decision-making process to get an optimal outcome. This is used in organizations where decisions are made following regulations to avoid risks.

2.4.2 Model-Driven Models

The model-driven model, grounded in bounded rationality and prospect theory, evaluates the governance and escalation pathways using a structured decision tree. It explains how incomplete information and loss aversion led to suboptimal policy trade-offs, reinforcing confusion across departments and delaying corrective action.

Model-driven decision-making models emphasize the use of established frameworks that guide structured reasoning and logical analysis. The rational theory explains a sequential process where decision-makers identify problems, develop alternatives, evaluate them objectively, and implement the most optimal solution. In contrast, the prospect theory explores how individuals evaluate risk and uncertainty, particularly in loss contexts. It suggests that decision-makers are more sensitive to potential losses than equivalent gains, a principle directly relevant to crisis management where reputational and financial losses dominate decision framing.

Model-based decision-making focuses on applying formal methodologies, structured processes, and analytic models to help inform decisions under conditions of uncertainty. Unlike cognition-driven approaches that emphasize heuristics and psychological biases, model-driven approaches use a system-based, systematic process to evaluate alternatives, consider risks, and optimize results.

These models support decision-makers by enabling the systematic quantification of uncertainty, the evaluation of trade-offs, and the anticipation of long-term consequences. Structured approaches such as decision trees, BN, and Markov Decision Processes (MDP) translate complex, uncertain scenarios into measurable and comparable components, thereby reducing ambiguity. Although uncertainty cannot be fully removed, these models provide a rational and defensible basis for selecting among alternatives, particularly in high-stakes contexts where intuitive judgement alone may be insufficient.

At the foundation of model-driven analysis is the rational decision-making framework, which follows a structured sequence: defining the problem, generating alternatives, evaluating options using evidence and probabilities, selecting the most favourable course of action, implementing the decision, and continuously monitoring and adjusting outcomes. While bounded rationality constrains perfect optimization in practice, this framework offers a disciplined and repeatable process that can be operationalized through analytical models. A central strength of the model-driven approach is its ability to impose structure and numerical rigor on complex decision-making situations, supporting more transparent and accountable choices.

However, this approach also has limitations. It requires accurate and sufficient data, technical expertise to interpret results, and consistent integration of model outputs into real-world decision contexts. If implemented without considering the human factors and organizational context, even the most sophisticated model can fail to influence actual decisions.

2.4.3 Model-Driven Decision-Making Process

The model-driven decision-making process applies structured analytical theories, specifically bounded rationality and prospect theory to explain how engineers and policymakers make choices under uncertainty. Bounded rationality asserts that decision-makers seek satisfactory rather than optimal outcomes because of cognitive, informational, and time limitations. In the context of large-scale infrastructure projects, this means engineers often rely on simplified evaluation criteria or organizational heuristics instead of exhaustive risk assessment. Prospect theory complements this view by showing that individuals evaluate outcomes relative to a reference point and are more sensitive to losses than equivalent gains. The model-driven process therefore provides a quantitative lens for understanding the systematic trade-offs embedded in engineering decision processes.

The process of developing, evaluating, and using mathematical, statistical, or computational models to predict possible potential future events is known as model-driven decision-making. Before choosing a course of action, the process assists decision-makers in weighing their alternatives by estimating risks and projecting the results of different scenarios.

The model-driven process as shown in Figure 2.3 provides insight into effective crisis decision-making under uncertainty. Firstly, it shows how data-informed observation can clarify problems and prospect framing can help organizations interpret crises not as failures but as opportunities for professionals to identify alternatives and determine consequences before making decisions.

The model-driven decision-making process follows a rational decision structure consisting of sequential steps: (1) problem definition, (2) identification of alternatives, (3) determination of consequences, (4) decision-making through optimal selection, (5) implementation, and (6) feedback and outcome assessment. Using the principles of expected utility, each alternative is assessed according to its potential benefits and associated risks. For example, a decision tree representation can quantify the expected utility of each choice based on probabilities of failure, maintenance costs,

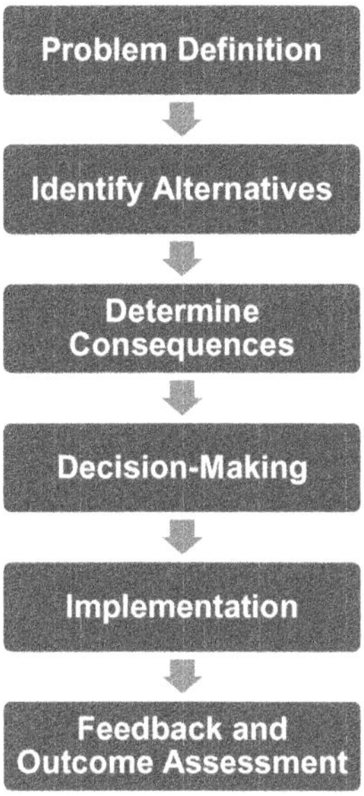

FIGURE 2.3 Model-Driven Decision-Making Process.

and construction time to implement a project. Applying prospect theory, decision-makers' tendency to weigh perceived gains (innovation, cost savings) more heavily than losses (future maintenance risk) can be modelled through a value function with diminishing sensitivity. The model structure integrates quantitative evaluation with behavioural insights, demonstrating how rational analysis combined with awareness of risk perception could enhance engineering decision quality.

2.4.4 Implications and Insights

Comparing the theoretical model with actual decisions reveals a significant gap between normative rationality and real-world behaviour. The model-driven process demonstrates that a balanced weighting of safety, cost, and aesthetics can favour a more conservative option. However, in practice, decision-makers operate under institutional and psychological pressures that can distort risk assessment. Economic incentives and political ambition shift attention towards short-term gains, while uncertainty can rationalize as acceptable innovation risk. The absence of formal utility-based or probabilistic analysis reflects the dominance of intuition over structured reasoning. If professionals execute a decision tree or Bayesian evaluation, it can reveal the high sensitivity of long-term safety outcomes and probabilities in uncertain situations. The contrast between theoretical predictions and actual choices underscores the necessity for embedding decision-analysis tools into engineering governance, ensuring that infrastructure decisions are informed by objective, data-supported evaluations rather than cognitive or political bias.

Complex situations with uncertainties are a good fit for the model-driven method because the executive-level decision is not made without structured processes, frameworks, and strategic planning. A model-driven lens emphasizes systematic decision-making, applying rational frameworks and established theories to evaluate how choices are made and what alternatives may produce better outcomes. Model-driven approaches offer organized frameworks for assessing organizational-level strategy and decision structure, while cognition-driven analysis concentrates on individual cognitive processes. The rational decision-making process, which proceeds logically from problem identification to execution, serves as the model's initial theoretical pillar.

In high-risk environments such as chemical plants, aviation control systems, or nuclear facilities, purely intuitive or heuristic-based decisions can be catastrophic. Structured models such as decision trees, BN, and MDP allow decision-makers to map complex scenarios into quantifiable components, thus reducing ambiguity. While these models cannot eliminate uncertainty, they provide rational grounds for choosing between alternatives when stakes are high.

A typically formed approach to decision-making, model-driven decision-making builds, assesses, and uses mathematical, statistical, and/or computational models to forecast future events. Before making a final decision, these models help evaluate options, estimate risks, and forecast the effects of many potential paths of study.

2.5 DATA-DRIVEN DECISION-MAKING

2.5.1 Theoretical Foundation

In theory and practice, data-driven approaches can counter the overreliance on subjective judgment that characterized early phases of incidents. However, the effectiveness of such an approach depends not only on data availability but also on the ability to integrate, analyze, and communicate findings transparently to decision-makers and the public.

Decision-makers must interpret a flood of data in real time, often under conditions of uncertainty, public scrutiny, and political pressure. While cognitive decision-making relies on intuition and experience, and model-driven frameworks (such as BN) structure uncertainty into probabilities, data-driven decision-making grounds actions directly in observed evidence. This approach leverages large volumes of information – sensor outputs, ecological monitoring, health records, and economic statistics to provide real-time feedback and guide adaptive responses.

2.5.2 Data-Driven Models

Data-driven decision-making model which focuses on use of BN, Markov chain, and other probabilistic reasoning techniques to model dependencies and calculate the likelihood of decision-making situations.

For example, Big Data Analytics and Monitoring Systems accommodate a data-driven model that can generate "big data" – high-volume, high-velocity, and high-variety datasets to reflect dynamic environmental and social conditions. In disaster contexts, these include satellite images, oceanographic sensors, weather models, and social media signals. The data-driven model posits that the greater the coverage and resolution of data, the stronger the ability to detect patterns and anticipate risks.

Another advanced data-driven model involves predictive and prescriptive analytics using artificial intelligence. By leveraging statistical and computational techniques, including time-series forecasting and machine learning algorithms, these data-driven models transform raw data into actionable insights that support judgemental decision-making. Predictive analytics identifies future outcomes (e.g., spill trajectories, ecosystem collapse), while prescriptive analytics suggests optimal interventions (e.g., where to place booms or prioritize cleanup). Structured analysis and quantitative evidence are the foundations of data-driven models. In terms of industrial applications, few of these consist of:

- Decision Tree and Probability Analysis: Assigning likelihood to decisions and results.
- BN: Probabilistic graphical models capture the conditional dependencies of risk factors. Bayesian reasoning is particularly useful in complex adaptive systems because it updates probabilities dynamically as new evidence arrives.
- Machine Learning Predictive Models: In large datasets, trends and anomalies can be found.

2.5.3 Data-Driven Process

The data-driven process as shown in Figure 2.4 explains how analytical intelligence can complement cognitive and theoretical models to create a holistic decision ecosystem. By bridging human judgement with quantitative insights, organizations can achieve hybrid decision intelligence – balancing intuition, logic, and empirical data.

The focus of data-driven decision-making is on supporting or even replacing human intuition with systematic, quantitative, and evidence-based approaches.

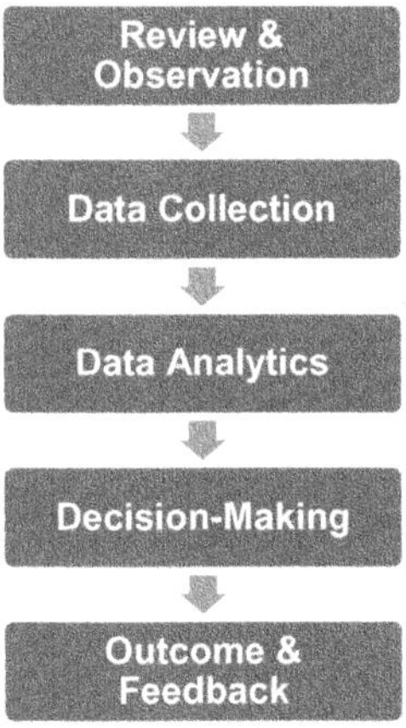

FIGURE 2.4 Data-Driven Process.

2.5.4 Implications and Insights

Unlike cognitive models that rely on experience and perception, data-driven methods turn unprocessed data into probabilistic knowledge. The ability to mathematically model uncertainty adds a crucial layer of analytical reliability in high-risk industries like defence and aviation. For example, the BN, a probabilistic graphical model that uses conditional probabilities to encode causal relationships between variables, is one of the most effective techniques in this category. BNs are frequently used in systems with incomplete or dynamically changing information for diagnostic reasoning, risk forecasting, and fault detection. In order to forecast the probability of a system failure or accident, they assist in integrating various data streams in aviation, including maintenance logs, environmental conditions, pilot performance metrics, and sensor readings.

Although cognitive and model-driven analysis take longer, data-driven models are transparent and objective-oriented. To properly interpret them, however, they need strong data systems and analytical literacy. Data-driven models can provide benefit from objectivity, transparency, and structure. They convert qualitative uncertainties into measurable risk scores using probabilistic reasoning. For example, a BN can assess how the combination of severe weather, poor maintenance, and flying in an urban area exponentially raises the risk of a crash.

Before taking any actual action, decision-makers can use these systems to see how risks spread. Nevertheless, there are two real-world limitations to the data-driven paradigm. In many military and developing-country contexts, where record-keeping is irregular and occasionally politically filtered, data integrity continues to be a persistent weakness. Inaccurate or incomplete maintenance logs make model outputs untrustworthy. Then, interpretability and human trust are critical factors that the data-driven models can assess. However, as example, operators or commanders in military operations may dismiss the model completely (algorithm aversion) or over-rely on it (automation bias) if they do not comprehend its logic. Therefore, even though data-driven reasoning improves analytical clarity, its usefulness is contingent upon institutional transparency, data maturity, and user cognitive acceptance.

The data-driven model explores evidence-based governance that emphasizes transparent, credible, and data-supported decision-making to enhance trust, legitimacy, and compliance. When professionals share consistent data openly, communities are more likely to accept both the risks and the policies. Together, these data-driven models provide a theoretical justification for embedding data systems into disaster response. Unlike purely cognitive or model-driven approaches, data-driven decision-making ensures that predictions are grounded in measurable reality and continuously updated as conditions evolve.

2.6 RATIONALE FOR MODEL SELECTION

Each model captures a distinct aspect of organizational decision-making, contributing data-driven insights through cognitive, strategic, ethical, and social dimensions. The cognition-driven model focuses on human judgement and psychological biases, the model-driven framework emphasizes structured theoretical reasoning, and the data-driven model adds an empirical, evidence-based dimension. Together, they create a comprehensive analytical triad that reflects how real-world decisions emerge from the interaction of intuition, logic, and data.

- Cognition-Driven Model: Select for its capacity to explain the pre-crisis strategic misjudgement that can lead to assess critical situations. This model uncovers the psychological mechanisms – bounded rationality, groupthink, and overconfidence bias that shape executive behaviour during the job execution. It demonstrates that even experienced managers can make flawed choices when cognitive shortcuts replace structured reasoning.
- Model-Driven Model: Choose to analyze critical situations, where crisis-response communication is important because it aligns with rational and prospect-based frameworks that guide structured evaluation under pressure. This model illustrates how professionals adopt a rational, empathetic approach by weighing potential risks and benefits, leading to the success of systems implementation.
- Data-Driven Model: Include to demonstrate the growing importance of analytical intelligence in contemporary decision-making. This model extends beyond cognitive and theoretical boundaries, showing how predictive and prescriptive analytics can transform the crisis response from reactive to proactive.

The cognition-driven model assumes that decision-makers act under uncertainty and cognitive constraints. Its strength lies in diagnosing human errors, but it cannot quantify their effect precisely. The model-driven approach, while systematic, assumes rationality and complete data conditions that rarely exist in crises. The data-driven model, though the most contemporary, depends on technological maturity and data quality, which may not always be accessible in fast-moving operational contexts.

Integrating these three models demonstrates how cognitive awareness, rational structure, and data intelligence complement each other in effective decision-making. The cognition-driven analysis explains why the failure occurred, the model-driven framework shows how recovery decisions were made, and the data-driven model proposes what could have prevented or improved these outcomes. This triangular analysis offers a holistic understanding of organizational decision-making under uncertainty.

2.7 CONCLUSION

In conclusion, the study of cognition-driven decision-making highlights the significant role of human factors such as biases, heuristics, and bounded rationality in shaping judgments under uncertainty. Cognitive tendencies like overconfidence, availability bias, and confirmation bias can distort risk perception and reinforce flawed decisions, as individuals often favour information that aligns with pre-existing beliefs. The principle of bounded rationality further suggests that decision-makers frequently satisfice rather than optimize, constrained by limited information, time, and cognitive capacity. Recognizing these limitations reframes decision-making as more than a purely rational or fact-based exercise and underscores the importance of structured tools and frameworks to support more reflective, balanced, and responsible decisions in complex engineering and high-stakes contexts.

Complementing cognition-driven insights, model-driven approaches offer a systematic means to reduce human error by replacing intuition with structured reasoning. By quantifying uncertainty and analyzing relationships between variables, these frameworks provide transparency and rigor, enabling organizations to identify optimal courses of action while adhering to regulations and mitigating risks. Meanwhile, data-driven models emphasize evidence-based governance, promoting decision-making that is transparent, credible, and supported by verifiable information. When data is shared openly and consistently, it strengthens trust, legitimacy, and compliance, ensuring that decisions are not only technically sound but also socially and ethically accepted. Together, cognition-, model-, and data-driven approaches form a comprehensive toolkit for navigating complex decision landscapes with accountability, insight, and resilience.

BIBLIOGRAPHY

Balmer, J. M. T., Powell, S. M., & Greyser, S. A. (2011). Explicating ethical corporate marketing. Insights from the BP deepwater horizon catastrophe: The ethical brand that exploded and then imploded. *Journal of Business Ethics*, 102(1), 1–14. https://doi.org/10.1007/s10551-011-0902-1.

Bazerman, M. H., & Moore, D. A. (2013). *Judgment in managerial decision making* (8th ed.). Wiley.

Calvi, G. M., Moratti, M., O'Reilly, G. J., Scattarreggia, N., Monteiro, R., Malomo, D., Calvi, P. M., & Pinho, R. (2018). Once upon a time in Italy: The tale of the Morandi Bridge. *Structural Engineering International*. Advance online publication. https://doi.org/10.1080/10168664.2018.1558033.

Festinger, L. (1957). *A Theory of Cognitive Dissonance*. Stanford University Press.

Kahneman, D. (2011). *Thinking, fast and slow*. Farrar, Straus and Giroux.

Kahneman, D., & Tversky, A. (1974). Judgment under uncertainty: Heuristics and biases. *Science*, 185(4157), 1124–1131. https://doi.org/10.1126/science.185.4157.1124.

Kahneman, D., & Tversky, A. (1979). Prospect theory: An analysis of decision under risk. *Econometrica*, 47(2), 263–291.

Klein, G. A. (1993). *Decision making in action: Models and methods*. Ablex Pub.

Moore, D., & Bazerman, M. (2013). *Judgment in managerial decision making* (8th ed.). Wiley.com. www.wiley.com/enus/.

Rania, N., Coppola, I., Martorana, F., & Migliorini, L. (2019). The collapse of the Morandi Bridge in Genoa on 14 August 2018: A collective traumatic event and its emotional impact linked to the place and loss of a symbol. *Sustainability*, 11(23), 6822. https://doi.org/10.3390/su11236822.

Simon, H. A. (1955). A behavioural model of rational choice. *Quarterly Journal of Economics*, 69(1), 99–118. https://doi.org/10.2307/1884852.

Tversky, A., & Kahneman, D. (1974). Judgment under uncertainty: Heuristics and biases. *Science (American Association for the Advancement of Science)*, 185(4157), 1124–1131. https://doi.org/10.1126/science.185.4157.1124.

Wickens, C. D. (1992). *Engineering psychology and human performance* (2nd ed.). HarperCollins Publishers.

3 Ethics and Socio-Cultural Values in Decision-Making

3.1 INTRODUCTION

In today's world, professional ethics and morale values play a critical role in taking decisions to execute complex tasks. Particularly, organizations are adopting emerging technologies that create uncertainties in decision-making processes. Professional bodies like engineers' institutions or computer society deliberately issue code of conduct to maintain ethical standards for a nation state. Every country has their distinct societal, economic, legal, and cultural identities that can create a unique system to shape moral values and professional code of conduct.

This chapter explores the relationships of morals with the cultural characteristics such as individualism vs. collectivism, uncertainty avoidance, and power distance. Further, we describe the virtuous feelings of a professional that engages in decision-making. People inherit their moral values from the family, society, peer groups, and digital interactions. These inherited morals influence their decision-making in every stage of life. People engage with academic institutions to acquire knowledge that can be applied for the benefit of society. In particular, professionals such as engineers, doctors, academics, and IT specialists play vital roles in translating this knowledge into responsible and impactful professional practice. However, professional codes of conduct can guide them to follow a standard principle in maintaining a harmonious environment within a society. In spite of social impact through moral values or peer groups, these external sources cannot influence professionals if they follow properly their professional codes of conduct.

The chapter develops an ethical and moral framework for engineers and IT professionals that includes human factors, ethical codes of conduct, morals, and virtues. The proposed framework explains a critical engagement and tension among multiple constructs to support a judgemental decision-making process in professional practices.

3.2 BACKGROUND ON ETHICS

In today's world, professional ethics and moral values play a vital role in taking decisions to execute complex tasks. When taking decisions in a group or individually,

DOI: 10.1201/9781003685449-3

preferences may vary considering contextual situations and availability of alternatives with ethical codes of conduct that make a trade-off between benefits and fairness. Others' actions and social influence can affect an individual's decision-making process. However, ethics and morals can play a critical role in making judgemental decisions and professional codes of ethics shape values such as modesty, honesty, impartiality, agreement with the rule of law, respect to cultural differences, trustworthiness, environmental support, and resource optimization. Professionals can follow organizational code of conduct to maintain their professionalism in delivering the highest standard of works to the society.

Ethics are all about perceived behavioural outcomes that are endorsed by universal norms across national boundaries that depend on people's inherited characters and moral values. People usually follow ethical viewpoints by applying moral values, decency, fairness, and virtuous thinking. By definition, ethics is defined as a standard protocol to show right or wrong behaviour. Professionals plan to take decisions by applying ethical code of practices, facts, ethical approaches, and consider the aftereffects on others.

Ethics may be understood as the rules of conduct that guide behaviour within a particular profession or cultural community, defining how tasks should be conducted responsibly, fairly, and in alignment with societal values. In this sense, ethics provides a framework for determining what constitutes professional and appropriate action. Within the field of information technology, ethical challenges emerge from a range of critical issues, including methods of data collection, the use and protection of personal information, data privacy and sharing practices, the proliferation of misinformation, personal integrity, identity theft, the deployment of artificial intelligence (AI), and the governance of complex data systems.

To address such challenges, professionals may draw on multiple approaches to ethical analysis when making decisions. Traditionally, ethical thought has been organized into philosophical classifications that distinguish between normative principles and codified rules. Normative or business ethics focus on values and expectations shaped by stakeholders, shareholders, and broader social influences, guiding what ought to be done in a given context. In contrast, codified ethics are expressed through formalized standards such as organizational policies, professional or industry codes of conduct, and legal and regulatory requirements, which collectively define the boundaries of acceptable practice.

In business ethics, the stockholder theory suggests that professionals in an organization have to maintain normative commitments to maximize benefits that co-create long-term benefits to the stockholders (Friedman, 2007). But they have to uphold their professional ethics and law by avoiding fraudulent or deceptive tasks for the organization. Therefore, this is a triangulation process to ensure every aspect from the point of stockholders' interest, professional ethics, and legal implications.

Conversely, professionals have to consider interests for associated stakeholders other than stockholders (company owners). Particularly, senior managerial professionals need to accommodate the requirements from different perspectives like customers, external vendors, suppliers, employees, regulators, and government authorities. Therefore, professionals like chief executive offices or managing directors are in a critical position to assure ethical concerns before making decisions.

However, this chapter addresses other considerable factors in explaining the ethical decision-making process, particularly for executing professional practices.

3.2.1 Professional and Organizational Ethics

Every professional group has their own distinct characteristics in deliberations to maintain professional integrities. Being a member of a professional association like Engineers Institution, Medical Doctors Association, Association of Chartered Professional Accountants, or Association of Computer Society, they have to abide by respective code of ethical conducts that create a standard protocol for the betterment of the society. Even though professionals follow international benchmarking practices, national laws, and organization values, professional ethics enforce them to maintain integrity, honesty, and professionalism. As a member of a professional body, each and every member must uphold their dignity, social respect, and accountability, which is articulated in the code of conduct. The code of conduct includes honesty, competence, public interest, knowledge development, and professionalism to co-create values to the society. Professionals can apply appropriate actions with ethical decision-making according to codes of conduct and claim those professional works at any instances.

Codes of ethics can guide professionals to follow a minimum level of standard in delivering their actions that are consistent with values and norms. There are common ethical considerations which can support professionals for judgemental decision-making. The notable ethical issues can be described as follows:

- Avoiding unprofessional behaviour and discrimination to others.
- Protecting data privacy and secrecy to maintain confidentiality.
- Uncompromising attitude with the quality of deliberations to meet deadlines.
- Assuring intellectual property (IP), security, and copyright for official documents.
- Protecting the environment for ensuring sustainability.
- Rigidness for bribery, corruption, and inducements.

Beyond professional ethics, organizations establish their own codes of ethics for employees, stakeholders, suppliers, vendors, and shareholders, creating binding obligations to uphold ethical business practices. These organizational ethics frameworks foster trust among all affiliated parties by promoting shared values, accountability, and motivation to act in accordance with responsible and ethical standards.

3.2.2 Ethical Viewpoints for Machine Intelligence

In the information age, both humans and machines increasingly integrate emerging technologies into their decision-making processes. Machines are evolving from traditional systems into intelligence-driven digital platforms that incorporate technologies such as AI, the Internet of Things, cloud computing, brain–computer interfaces, and digital twins, enabling the co-creation of value through machine intelligence. But there is a critical consideration in the decision-making process, because it can deliver hallucinated information that can create uneven consequences in job deliberations. That is why we need to address the ethical commands for machines to get optimum outcomes using machine intelligence. For example, regulators and governments can formulate new policies and guidelines for autonomous driving rules and intelligent traffic systems to acknowledge the advanced technologies like AI, IoT, smart city, and autonomous driving in adopting ethical viewpoints for the betterment of the society. In other words, we can think about the smart logistics, where respective regulators can develop a new protocol in addressing the inclusion of robots, IoT, digital twins, and smart contract.

3.2.3 Socio-Cultural Factors

3.2.3.1 Social Norms

Social impacts are crucial in the decision-making process, particularly for professionals to execute project activities or tasks. In the earlier stage, people develop their social norms through interactions with peer groups, families, friends, and office colleagues (Cialdini & Trost, 1998). These perceived social norms influence people's behavioural responses through cognitive engagements. In the present world, we can connect the world using internet, and it allows us to gather valuable information. Particularly, social network sites (SNS) connect people from the distant parts of the world that can engage more visibilities and interactions with ease of access. Nowadays, digital platforms create more enabling environment for interactive engagements, specifically for professionals in making their decisions. In the age of AI, professionals can extract valuable insights for their decision-making using generative AI. In contrast, organizations can accommodate agentic AI in supporting large volume of customers with a limited number of professionals. All those interactions and engagements influence peoples' social norms in the digital era.

Like peer influence by human, interactions with advanced technologies can also impact in a negative way such as defamation, misinformation, hallucination, hacking, fraud, scamming, breach of privacy, leakage of confidentiality, and network security that can create adverse sentiment towards technologies or mislead people in their decision-making process. Therefore, digital platforms or other information systems can play a critical role in perceiving social norms that subsequently influence peoples' decision-making. This is an emerging issue for professional institutions to revisit their professional code of conduct addressing the impacts of advanced technologies.

3.2.3.2 Morals

People inherit moral values from the societal system. It develops their core beliefs about something being right or wrong by an individual. The moral can be defined as a core habit or principle of an individual in respect to right or improper conduct in executing activities. Individual moral values can create a harmonious society that enables to maintain optimum beneficence, less harm, respect for autonomy, and justice in a society.

3.2.3.3 Cultural Influence

Culture is the collective programming of people that recognizes their characteristics in the decision-making process. National and organization culture develops people's morals and ethics that influence their choices and preferences through decision-making. Cultural characteristics such as individualism, power distance, and uncertainty avoidance are critical factors in developing people's cognitive feeling or experience to adopt technologies (Hofstede, 2011). Deviation across diverse cultural characteristics lead to changing perceptions and attitudes to the conduct in which the decision-making is developed.

For instance, cultures with low uncertainty avoidance have lower experiences of considering project risk than cultures with higher uncertainty avoidance. Therefore, culture can influence professional's decision-making, particularly for executing engineering practices. Considering organizational culture, uncertainty avoidance acts as a significant factor in determining how professionals can potentially adopt technological systems. Conversely, effective adoption of IT is more likely to happen in a low power-distance environment that clearly indicates the influences of cultural values in organization decision-making. An information system is designed to accommodate professionals' autonomy among operational activities in one subunit of an organization located in a low-power distance culture being transferred to a different subunit inhabiting in a high-power distance culture.

3.2.3.4 Legal Implications

Countries have formulated specific laws related to ICT and other compliance of digital services. Every nation-state formulates their laws considering industrial contexts, including specific laws related to information security, data privacy/protection, and cyber surveillance that confirm citizen rights with ethical values. Special laws such as contract law, consumer protection law, data privacy and protection law, IP law, and criminal law are directly associated with the ethical practices, particularly for ICT and engineering sectors. For example, contract law supports a business entity and customers to make enforceable agreement that protect rights and ethical obligations. These obligations guarantee customers' rights, service assurance, and protect from misleading conduct.

In contrast, IP and copyright acts assure IP assignment and licensing to protect creative and confidential works from unauthorized users. Furthermore, privacy law can regulate with dealing personal identifiable and sensitive information. In addition to specific law, regulators issue compliance directives, protocols, and standard regulations to maintain business activities in an ethical way.

3.3 ETHICAL FRAMEWORK

3.3.1 Ethical Approaches for Decision-Making

Professionals can follow various approaches in analyzing ethical values before making decisions. The decision-making process is critical in nature, even if it follows guidelines or approaches. Professionals need to address multiple factors such as organizational rules, social impacts, government and regulatory policies/framework, cultural associations, professional affiliations, legal implications, and contextual awareness.

Approach I shows a goal–duty–right–centred ethical analysis that includes stakeholder analysis to ensure optimum gains for the associated stakeholders. Professionals can extend their analysis through classifications, interest-power grid breakdown, priority assessment, and impact measurement. Then, they can perform critical evaluations to identify the best possible option from available alternatives that can avoid any violations of professional codes of conduct with the assurance of optimum benefits to the society. Finally, professionals need to protect human rights and confirm legal obligations in executing their tasks.

Figure 3.1 illustrates a process flow map that explains three stages for ethical decision-making. Firstly, the goal-centred analysis allows professionals to examine how an action can support in getting maximum level of gains in the workflow of any particular task. Secondly, the duty-centred analysis can make an understanding about ethical duties and to assess the best alternative for a proper execution. In the third stage, professionals can extend their analysis to ensure human rights, legal implications, and regulatory compliances.

In this approach to analyzing ethical situations, professionals can follow three structured stages to examine key considerations that provide valuable insights into what is right, what goals should be pursued, and what duties must be fulfilled. This process supports ethical and logical action, enabling professionals to demonstrate responsibility, integrity, and professionalism in their service to society.

Approach II allows professionals to identify probable answers to 13 questions that can explore the ethical standards by extending sentiment analysis and accommodating rating for decision-making (Burls et al., 2011). These questionnaires particularly assess professional's perceived feeling in executing tasks. Table 3.1 shows the definition and types of questionnaires along with expected sentiment and associated

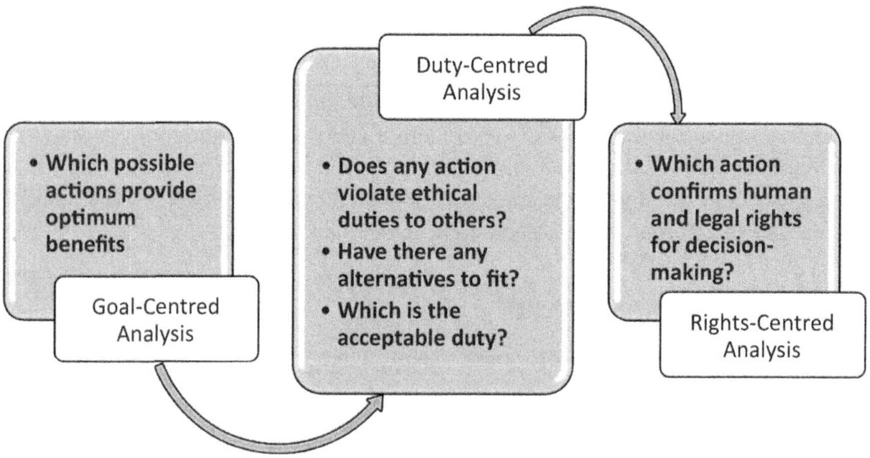

FIGURE 3.1 Goal–Duty–Right–Centred Ethical Approach.

TABLE 3.1
Thirteen Questionnaire–Centred Ethical Approach

Questionnaire Type	Questionnaire (Context)	Expected Sentiment	Rating for Decision-Making (Low-Medium-High)
Close-Ended	Is it legally viable?	Yes	High
Open-Ended	How do you assess the situation (anxious, fearful, etc)	Positive	Medium
Close-Ended	Does your moral interrupt you?	No	High
Close-Ended	Any contradictions with existing laws	No	High
Close-Ended	Consistency with previous practices	Yes	Medium
Open-Ended	Consequences of being appeared in the news (print, electronic, or social media)	Positive	Medium
Close-Ended	Integrity with disclosure (Lie) about the process	No	High
Close-Ended	Uneven situation that requires critical response	No	Medium
Close-Ended	Fairness in performing activities	Yes	High
Close-Ended	Have there any scope to discuss the action with seniors?	Yes	Medium
Close-Ended	Could you recommend to your peer groups?	Yes	Low
Close-Ended	How would you react if it relates to your closest one?	Positive	Medium
Close-Ended	Any obligations to keep secret for disclosing actions	No	Medium

rating values. The outcomes of the feedback guide professionals in their ethical decision-making process.

In critical situations, Approach III can be applied through four stages to explore the forward-looking decisions that can mitigate ethical dilemmas. Particularly, professionals can execute the four-stage decision-making approach to avoid reputational and ethical risks. This approach supports to extend a thorough analytical process for current critical situations, as well as providing a precedence for future events. Figure 3.2 illustrates the process flow in examining the situational facts, legal implications, ethical propositions, resolution mechanisms, and policy frameworks.

3.3.2 Ethical Decision-Making Framework

The figure (as shown in Figure 3.3) exhibits a four-stage ethical decision-making framework that leads professionals from understanding to identify a situation towards a final evaluation. The framework is organized into four stages such as preliminary, execution, examination, and final stage (including post-decision analysis) in the

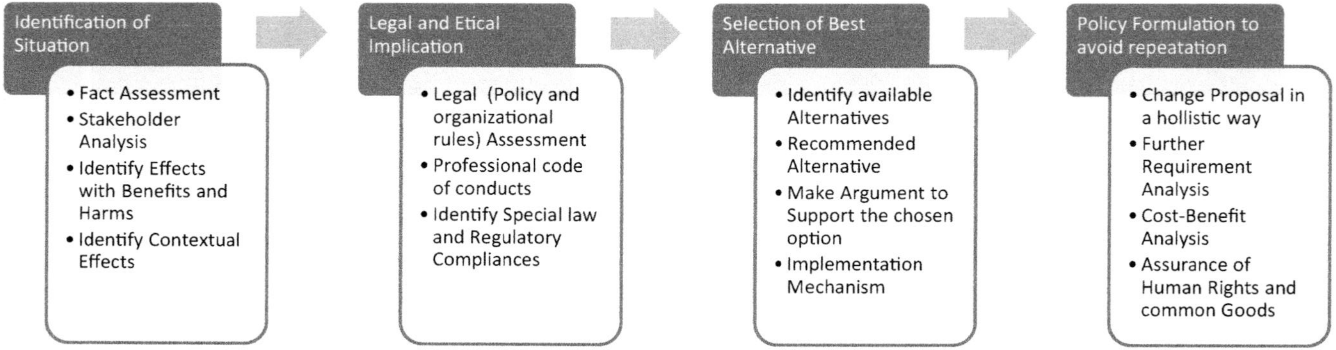

FIGURE 3.2 Four-Stage Ethical Decision-Making Approach.

decision-making process. Each stage includes specific activities, showing how contextual understanding, cultural awareness, analytical tools, and evaluation processes interact to support an informed ethical decision.

In the preliminary stage, the emphasis is on understanding the broader environment of the problem. This includes assessing contextual factors, reviewing professional and organizational codes of conduct, and recognizing national, organizational, social, legal, and cultural influences. This stage ensures that the decision-making process begins with a comprehensive situational awareness and ethical grounding.

The execution and examination stages shift towards methodological structure – selecting relevant ethical codes, identifying cultural values, applying socio-legal considerations, and using appropriate questionnaires or analytical tools. Here, professionals operationalize the assessment by building constructs, choosing methods, analyzing results, and comparing alternatives. This stage acts as the analytical core of the framework.

Finally, the decision-making stage consolidates outcomes into actionable recommendations. It emphasizes planning based on analyzed results and includes post-decision reflection, which is essential for organizational learning and improving future decision processes. Overall, the figure illustrates a systematic, culturally aware, ethics-driven pathway for making justified professional decisions.

3.4 IMPLICATIONS AND DISCUSSION

In contemporary professional practice, ethical principles and moral values play a decisive role in shaping decisions related to complex and high-stakes tasks. Whether decisions are made individually or collectively, preferences often diverge due to contextual constraints, competing alternatives, and differing interpretations of ethical codes, requiring deliberate trade-offs between efficiency, benefit, and fairness. Social influence and the observed actions of others further shape judgment, introducing implicit pressures that can amplify or constrain ethical reasoning. The implication for practice is clear: decision-making cannot be treated as a purely technical exercise but must be understood as a socially embedded process in which values, norms, and professional identity continuously interact.

Distinct professional communities develop their own deliberative norms to preserve integrity and public trust, with codes of ethics providing a shared baseline for acceptable conduct. These codes do not prescribe every action but establish minimum standards that align professional behaviour with societal values and institutional expectations. In the information age, this ethical landscape is further complicated by the growing integration of intelligent digital systems into decision processes. As machines evolve into intelligence-driven platforms leveraging AI, the Internet of Things, cloud infrastructures, brain–computer interfaces, and digital twins, they increasingly participate in value co-creation alongside humans. However, the potential for biased outputs or hallucinated information introduces new ethical risks, underscoring the need for clearly articulated ethical commands and governance mechanisms to ensure that machine intelligence supports, rather than undermines, responsible professional judgment.

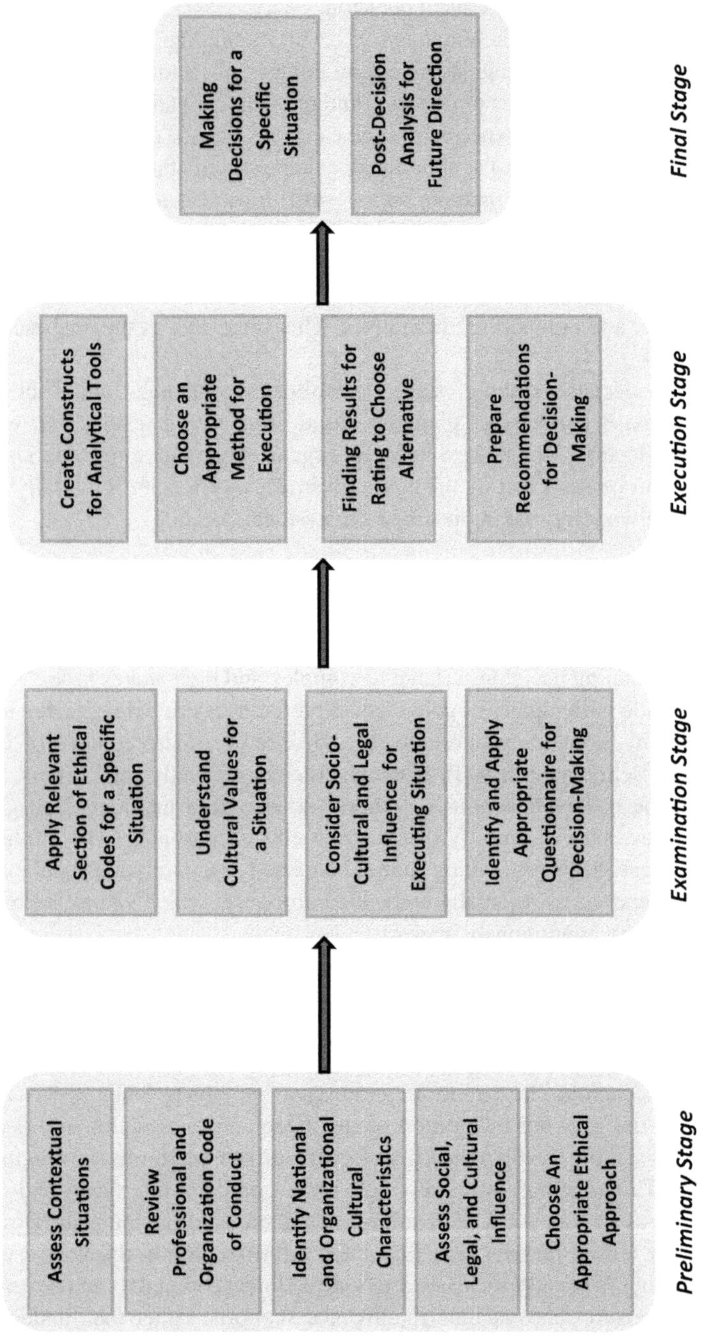

FIGURE 3.3 Ethical Decision-Making Framework.

Against this backdrop, digital platforms and information systems are not merely tools but active mediators of social norms that influence human decision-making. This development calls for professional institutions to critically revisit and update their codes of conduct in light of advanced technologies and cultural contexts, recognizing culture as a collective programming that shapes how ethical choices are perceived and enacted. The chapter responds to this need by discussing three complementary ethical approaches: a goal–duty–rights–centred analysis that incorporates stakeholder evaluation to balance competing interests; a 13-question, sentiment-informed approach that supports reflective and comparative ethical assessment; and a forward-looking, four-stage approach designed to anticipate and mitigate ethical dilemmas. These approaches culminate in a four-stage ethical decision-making framework – preliminary analysis, execution, examination, and post-decision evaluation – offering professionals a structured pathway from ethical awareness to accountable action and continuous learning.

BIBLIOGRAPHY

Burls, A., Caron, L., Cleret de Langavant, G., Dondorp, W., Harstall, C., Pathak-Sen, E., & Hofmann, B. (2011). Tackling ethical issues in health technology assessment: A proposed framework. *International Journal of Technology Assessment in Health Care*, 27(3), 230–237. https://doi.org/10.1017/S0266462311000250

Cascio, W. F., & Montealegre, R. (2016). How technology is changing work and organizations. *Annual Review of Organizational Psychology and Organizational Behavior*, 3(1), 349–375. https://doi.org/10.1146/annurev-orgpsych-041015-062352

Cialdini, R. B., & Trost, M. R. (1998). Social influence: Social norms, conformity, and compliance. In D. T. Gilbert, S. T. Fiske, & G. Lindzey (Eds.), *The Handbook of Social Psychology* (4th ed., Vol. 2, pp. 151–192). McGraw-Hill.

Fishbein, M., & Ajzen, I. (1975). *Belief, attitude, intention and behaviour: An introduction to theory and research.* Addison-Wisely.

Friedman, M. (2007). The social responsibility of business is to increase its profits. In W. C. Zimmerli, M. Holzinger, & K. Richter (Eds.), *Corporate Ethics and Corporate Governance* (pp. 173–178). Springer. https://doi.org/10.1007/978-3-540-70818-6_14

Hofstede, G. (2011). Dimensionalizing cultures: The Hofstede model in context. *Online Readings in Psychology and Culture*, 2(1), 8. https://doi.org/10.9707/2307-0919.1014

Leidner, D. E., & Kayworth, T. (2006). Review: A review of culture in information systems research: Toward a theory of information technology culture conflict. *MISQ*, 30(2), 357–399. https://doi.org/10.2307/25148735

Smith, H. J. (2002). Ethics and information systems: Resolving the quandries. *Database for Advances in Information Systems*, 33(3), 8–22. https://doi.org/10.1145/568425.568444

4 Human Factors, Situation Awareness, Critical Thinking, and Value Co-Creation in Decision-Making

4.1 INTRODUCTION

In decision-making, solutions are often developed through multiple stages under conditions of uncertainty, where human factors play a decisive role in shaping judgment. Understanding these human factors – both as individuals and as professionals is essential for executing tasks effectively. While it is commonly assumed that academic knowledge and industrial experience are sufficient to manage critical responsibilities, systems engineering has demonstrated that explicit consideration of human factors is indispensable when making high-stakes professional decisions.

In many contexts, professionals must continuously adapt to uneven and rapidly changing situations that demand new forms of deliberation or immediate action, particularly when safeguarding information systems. Such moments evaluate professional capability and resilience. Situational awareness has therefore emerged as a key concept, enabling professionals to develop confidence and expertise across diverse scenarios while maintaining ethical standards and public safety. This is especially relevant in environments such as preventing aircraft collisions, managing complex adaptive systems, or coordinating large populations during the establishment of critical infrastructure projects like nuclear power plants. Consequently, when implementing information systems, it is essential to anticipate and evaluate the full range of uncertainties that may influence decision-making processes.

How can we maintain a value chain in the decision-making process? One of the most important techniques is to build capacity in acquiring a critical thinking skill. This can boost professionals' confidence in taking any sort of challenges, and they feel comfortable in taking critical decisions applying their gained experiences and methodological approaches. Critical thinking is not only a skill, but also a rare type of quality that professionals can get to execute any uneven situation without compromising qualities. In another setting, professionals can develop their versatile knowledge from various sources, and they can innovate latest ideas or engage emerging technologies to co-create values for their organization. The value co-creation through

DOI: 10.1201/9781003685449-4

technological adoption can leverage professionals in taking judgemental decisions, and it even supports them to be more transparent and accountable to the society.

4.2 VALUE CHAIN IN DECISION-MAKING

4.2.1 Human Factors

Professional decisions are challenging due to complexity and uncertainty. Human factors such as moral values, ethics, and emotions also influence outcomes in taking decisions. Professionals must be aware of human factors that can impact their decision-making processes and enhance their capabilities to tackle critical situations. There are few factors such as cognitive, organizational, social, and environmental factors, that have a strong influence on people's decision-making processes. For example, peer influence by friends can influence professionals in taking decisions to select a vendor to implement and execute an information system. It is a critical skill in addressing those human factors as an art of professionalism that comes with great responsibilities.

Human factors engineering and ergonomics surfaced in the 20th century to tackle real-world challenges involving system performance and safety. The field applies a scientific lens to understand human behaviour, emphasizing systematic patterns rather than individual traits such as genetics or cultural background. Human factors are examined as integral elements within complex adaptive systems, shaped by interacting cognitive, social, and environmental influences (Wickens et al., 2004). Human factors research recognizes decision-making as a systems-based cognitive process shaped by multiple interacting influences rather than individual judgement alone.

Human factors can be defined as a construct that includes cognitive, social, organizational, and environmental influences to shape how people process information in their cognitive minds, judge alternatives, and take an initiative to perform actions. In professional and engineering contexts, where decisions often involve extreme consequences for safety and ethical risks, these human factors play a vital role in delivering judgemental decision-making.

4.2.1.1 Cognitive Factor

Cognitive factors pertain to the mental mechanisms through which the human brain processes information. Information processing encompasses the sequential cognitive operations that transform sensory inputs into meaningful perceptions, decisions, and actions. As information enters through the sensory channels, it is rapidly screened by perceptual systems that identify patterns, prioritize salient cues, and filter out extraneous stimuli. This initial phase is governed by attentional processes, which function as selective gatekeepers determining which inputs advance to conscious cognition. Attended information is then retained in working memory, a capacity-limited system responsible for maintaining and manipulating transient data such as numerical values, visual features, or task instructions. Within this stage, new sensory input is integrated with pre-existing knowledge stored in long-term memory, enabling the formation of coherent interpretations, predictive judgements, and mental models.

Cognitive factors play a cardinal role in determining how professionals process information, evaluate available alternatives, and choose the best option to perform a specific task. As cognitive mechanism unfolds, the brain involves a range of functions including inference, pattern recognition, and cognitive simulation to identify alternatives and give a direction towards decision-making. These brain functions are processed by neural networks within the prefrontal cortex, which support professionals to perform tasks with complex reasoning, integration of multiple information, and planning. At the same time, associative cortical regions consider current situational information cues with perceived experiences, allowing professionals to make a consensus in delivering judgemental and experience-based decision-making.

Although decision-making often occurs instantly, it engages the concurrent activities of multiple neural pathways that maintain an equilibrium to present intuitive and rational decisions with analytical reasoning. However, cognitive load imposes limits on processing capacity; when the volume or complexity of information exceeds the bounds of working memory, the quality of decision-making may decline or deteriorate. Under critical uncertainties, professionals may make irrational decisions when faced with inadequate or excessive information, or when influenced by misleading or fabricated outputs generated by machine intelligence, such as generative artificial intelligence (AI). Under such constraints, professionals may depend on heuristics, ethical code of practices, perceived experiences, and decision-making frameworks to avoid critical loads as the brain utilizes less cognitive resources to make rational decisions.

Overall, cognitive factor deals with a dynamic process in which memory loads, perceived experiences, and analytical reasoning constantly interact inside the brain. Information is not simply sensed but actively processed in the brain that interprets sensory signals considering expectations, previous knowledge, and available source of information. This natural function rationalizes why professionals and engineers may observe the same set of data in a unique way depending on their experience, anxiety, situational context, or cognitive biases. Understanding these processes facilitates to clarify how decisions are made within the mind and why professionals' decision-making is both deep and susceptible in executing complex adaptive systems.

In complex engineering endeavours such as information system design, autonomous vehicle development, drone-enabled smart logistics, robotic surgery, and intelligent traffic system implementation, cognitive overload can disrupt decision-making and lead professionals to overlook critical information arising from system interactions. Sustaining situational awareness and employing structured decision-support methods (e.g., fault tree analysis) can help mitigate these risks. Human factors are not only related to cognitive functions, but also other external factors influence professionals' decision-making.

4.2.1.2 Social and Organizational Factor

Social and organizational influences play a crucial role in shaping professionals' human factors, requiring careful consideration of external pressures and peer dynamics. As social beings, professionals are influenced by peer groups such as family and friends, as well as by perceived social norms rooted in cultural values and societal expectations. These social factors shape cognitive development and decision-making processes, often varying across geographical and cultural contexts. Over time, individual norms and virtues become acculturated through social influence, forming external forces that significantly affect cognitive judgment and professional decision-making.

Organizational culture, including shared values, principles, and ethical standards, plays a pivotal role in shaping how professionals approach and execute decisions. Each organization establishes a distinct cultural environment that influences how professionals interpret their responsibilities and align their actions with the organization's strategic value chain. Effective assimilation into this culture can significantly impact professionals' critical reasoning, particularly when navigating complex or uncertain decision-making scenarios. These organizational dynamics shape not only individual judgement but also team-based decision processes, affecting the organization's credibility and accountability to society. In this context, ethnographic workplace analysis provides valuable insights by examining real work practices and informing the design of socio-technical systems that better support professional decision-making.

Engineering decisions are shaped by organizational environments that foster effective communication, shared mental models, and psychologically safe mechanisms for reporting concerns. High-reliability organizations such as those in aviation, nuclear operations, and medical device manufacturing intentionally train teams to reduce hierarchical barriers and use structured communication protocols. While some organizations rely on a top-down decision-making approach, others adopt a bottom-up model that actively solicits feedback and insights from staff to enhance the quality and relevance of decisions.

Professional decision-making seldom occurs in isolation; rather, it is shaped by the quality of communication within teams. Ambiguous attitudes or poorly communicated viewpoints can lead professionals, particularly in multidisciplinary engineering groups to form conflicting assumptions that compromise judgemental and rational decision-making. Organizational culture further conditions these dynamics by either fostering collaborative engagement or, conversely, discouraging open dialogue in ways that may heighten interpersonal conflict or reduce the quality of interactions with external clients.

Authority gradients also play a critical role in shaping the decision-making process. In high power-distance contexts, hierarchical structures often legitimize top-down directives, reinforcing deference to senior positions. By contrast, professionals working within low power-distance cultures tend to adopt more horizontal organizational forms that promote shared responsibility, mutual respect, and collective decision-making.

4.2.1.3 Environmental Factor

Environmental conditions play a pivotal role in shaping decision-making within engineering practice. Professionals must account for environmental impacts, public expectations, and broader societal consequences when translating technical activities into real-world outcomes. Their capacity to demonstrate professional competence often depends on their ability to align organizational priorities with environmental

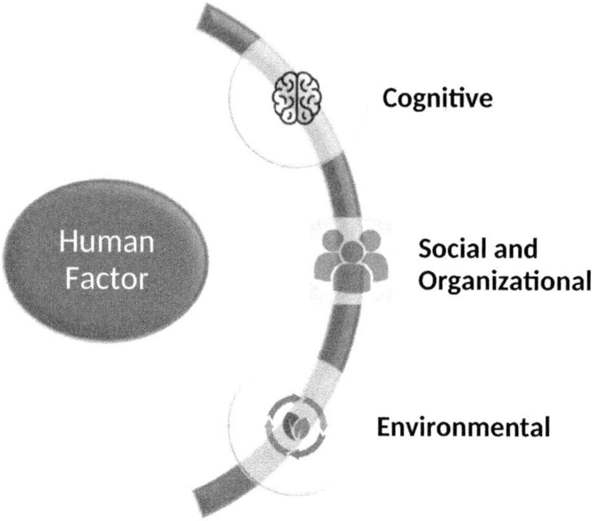

FIGURE 4.1 Elements of Human Factor in Decision-Making.

considerations, thereby securing stakeholder confidence in the implementation of information systems that maximize operational performance.

In undertaking complex decisions, IT professionals and engineers routinely evaluate design specifications, workforce capacity, ethical and environmental risks, commissioning requirements, quality assurance procedures, and regulatory implications. Such assessments support the resolution of critical uncertainties and enhance decision robustness. Attention to ergonomic principles and human-centred design (HCD) further strengthens professional judgement by ensuring that systems are aligned with human capabilities and limitations. Environmental influences including fatigue, weather conditions, and natural hazards also shape human factors by affecting decision accuracy and situational awareness. Moreover, engineers must anticipate the implications of human–machine interaction, particularly interface design and contextual environmental constraints, which determine how technologies are adopted and used in practice.

Therefore, human factor not only includes professionals' cognitive functions, but also considers peripheral factors like social, organizational, and environmental factors that influence their critical decision-making in complex adaptive systems. Human factors in decision-making as shown in Figure 4.1 embeds professionals' cognitive engagements with social and organizational impacts.

4.2.2 Situational Awareness

Although IT professionals and engineers possess substantial theoretical and practical expertise, they inevitably encounter critical situations in which decision-making requires more than technical proficiency. In such moments, the ability to exercise situational awareness becomes essential. This helps to develop awareness about associated risks and let professionals to navigate uncertainty, anticipate emerging risks, and respond effectively to perform critical activities. This process allows professionals to develop qualities for dynamic conditions that distinguish them exceptionally from those who rely solely on routine knowledge (Endsley, 1995).

When situational awareness is insufficient, the likelihood of unethical behaviour increases. Diminished attention and poor judgement can fuel self-serving cognition, self-deception, and unconscious biases. In this sense, a lack of mindfulness does not merely lead to technical oversights; it can also create conditions in which ethically questionable actions appear more common or difficult to interpret.

Enhancing the effectiveness of complex adaptive systems requires situational awareness to be maintained at every level of operation. Professionals must carefully observe and monitor human–machine interactions, recognizing both the capabilities and limitations of the systems they oversee. They must also appreciate how surrounding elements such as people, technologies, and environmental uncertainties shape system behaviour. Furthermore, understanding the current and potential future actions of information systems is essential, as these actions may directly influence organizational objectives or mission outcomes.

While understanding the historical context of a problem is indispensable, modern professionals frequently confront overwhelming volumes of data. Visual analytics has emerged as a vital approach for synthesizing and interpreting such complexity in making decisions. Visual analytics offers a means to extract timely, meaningful, and interpretable insights. It enables both the confirmation of expected patterns and the discovery of unforeseen relationships, thereby supporting more informed and strategic decision-making.

Professionals must continuously refine their capabilities, particularly by cultivating strong situational awareness, while also guarding against becoming "bad apples" – a term often used to describe individuals whose lack of competence undermines their performance and professional integrity. Competency gaps can even affect seasoned experts when they fail to recognize or adapt to the demands of their role. To counter this, professionals should develop a deliberate awareness of their surroundings, actively interpret what unfolding events mean in the present, and anticipate how these factors may influence future outcomes. Through this ongoing attentiveness, they can strengthen both their judgement and their ability to act responsibly within complex environments.

The quality of a professional's situational awareness is profoundly shaped by the design of the system they interact with. System functions, inherent limitations, onboard technologies, and mechanical characteristics all influence how effectively professionals can perceive relevant cues, comprehend their significance, and project future states. There are three components such as perception, comprehension, and projection that form the foundation of situational awareness and determine how well professionals can navigate complex operational environments.

4.2.3 Critical Thinking

Critical thinking is a disciplined, methodological process that begins with the clear identification of a problem. From this point, professionals engage in critical reading to examine relevant information and assess its credibility. The approach then progresses to a deeper analytical phase in which gaps are recognized, assumptions are questioned, and potential alternatives are explored through innovative ideation. Such thinking is indispensable in professional practice, particularly when decisions must be made under conditions of uncertainty or when solutions are not readily evident. Establishing well-defined objectives at the outset enables professionals to extend their methodological inquiry in ways that consider contextual factors and factual precision.

In fields such as information systems implementation and product development, the careful analysis of facts is essential to understanding the nature of the problem. Professionals frequently employ triangulation process, drawing on multiple data sources or methodological approaches to verify and validate evidence, thereby strengthening the authenticity and rigour of their arguments. The critical thinking process involves systematically sorting, organizing, and evaluating facts and information in order to articulate the problem clearly and devise effective, contextually grounded solutions (Butler, 2012).

Furthermore, critical thinking encompasses collaborative intellectual engagement. It requires professionals to articulate and evaluate differing viewpoints, identify the strengths and limitations of others' claims, examine evidence from multiple perspectives, and balance competing arguments with fairness. It also involves recognizing flawed logic and persuasive biases while maintaining perseverance throughout the research and reasoning process. This reflective and analytical practice enables professionals to synthesize existing knowledge and generate new insights or innovative solutions that contribute meaningfully to society.

BOX 4.1 EXAMPLE SCENARIO IN CRITICAL THINKING

What distinguishes traditional newspaper reading from an academic literature review grounded in critical thinking?

Traditional newspaper reading is largely oriented towards gathering information. Readers engage with news articles to keep abreast of national and international events, gaining peripheral awareness of current issues, public debates, and factual developments. The objective is primarily informational: to understand what is happening and why it matters in the broader societal context.

In contrast, an academic literature review serves a fundamentally analytical and investigative purpose. Researchers and scholars examine journal articles, book chapters, and empirical studies not merely to absorb information but to interrogate it. They seek to identify research gaps, conceptual limitations, methodological weaknesses, and unresolved problems within the existing body of knowledge. Through critical evaluation, they explore potential solutions, formulate new questions, and generate innovative ideas that advance scholarly understanding.

For this reason, critical thinking becomes an essential competency for professionals and researchers. It enables them to move beyond passive consumption of information towards the active construction of knowledge, ultimately supporting rational, evidence-informed decision-making in both academic and professional domains.

4.2.4 Value Co-Creation

Value co-creation represents a holistic and collaborative endeavour in which multiple stakeholders such as clients, professionals, platform designers, and technology developers work collectively to enhance user experience within personalized digital environments. In the contemporary digital era, emerging technologies such as AI, virtual and augmented reality (VR/AR), the Internet of Things (IoT), cloud computing, 5G networks, and digital twins converge to generate co-created value through interactive and immersive experiences tailored to user needs and preferences. For instance, the fusion of physical reality with virtual spaces through VR allows platforms to cultivate enriched experiential value that strengthens user engagement.

Professionals across sectors can draw on these technologies to co-create value in ways that reinforce brand equity, support online brand communities, and deepen consumer engagement and retention. The evolution of user experience is increasingly linked to the integration of advanced digital tools that enable users to perceive and participate in co-created value within public and organizational settings. By incorporating AI-driven, adaptive recommendation systems, professionals can offer personalized and responsive interactions that heighten the sense of co-produced value across diverse industries. As digital transformation accelerates, value co-creation has become a pivotal consideration in decision-making, particularly for engineers and IT practitioners who operate within complex, interactive digital ecosystems.

In the information age, professionals need to deal with both machine and human, and machine intelligence can also influence their decision-making. For example, AI chatbot using generative and agentic AI can guide professionals in taking critical decisions that can have an impact on value propositions for an organization. Similarly, HCD process is an example scenario, where value co-creation is critical in decision-making.

BOX 4.2 DECISION-MAKING IN HUMAN-CENTRED DESIGN PROCESS

Human-centred design (HCD) incorporates a close collaborative value co-creation aspect that explores user requirements, task dependencies, environments, contextual expectations, technological impacts, and socio-cultural values. Contextual design process involves designing products based on the professional's understanding of delivering co-created values to customers for ensuring optimum level of their satisfaction. It focuses on customer retention process for business continuity.

Improving value co-creation through sensory information cues, enhancing customer service deliberations, interactive feedback, and facilitating data analytics can support professionals in developing strategies and recommendations for better decision-making.

4.3 INTEGRATED DECISION-MAKING FRAMEWORK

Decision-making emerges from the integration of cognitive, organizational governance, social values, and environmental impacts. These realms overlap and induce each other, leading to oversights that typically develop not only from professionals' failures but also from system-level inefficiencies.

This chapter explains an integrated decision-making framework to guide professionals on how they can follow sequential steps in making judgemental decisions. At the initial step, cognitive factors including perception, attention, memory, cognitive load, information processing, and emotional intelligence regulate how professionals understand insights and choose an action. Decision-making is further influenced by social values, including social norms, cultural values, interpersonal trust, social influence, authority gradients, hierarchy, and power distance that affect the flow of making decisions and the capacity to take challenges as a professional in complex situations. However, these cognitive and social processes are surrounded with organizational structures such as governance, leadership, policy, competency development, safety culture, psychological safety, and resource constraints, all of which form the decision-making. Furthermore, environmental factors such as weather, system design, regulations, operational constraints, fatigue, and resistance also directly affect human factors and error susceptibility in taking decisions.

Professionals can concentrate on how to co-create values in implementing information system or in executing product development by assessing few factors such as emerging technology adoption, brand equity, online communities, new service provisions, deep engagement with customers, and retention process for ensuring business sustainability. Then, professionals can gain skills on situational awareness to tackle uncertainties or critical situations. The situational awareness can be built through capacity development, anticipation, risk assessment, and recall past experiences that can allow them in making rational decisions instantly without compromising quality of works. Then, they can apply critical thinking process to identify gaps and explore alternatives with innovative ideas to execute their tasks. Lastly, professionals can make decisions considering those sequential steps that enable them to be more rational and judicious in their professional tasks.

Together, these interdependent domains form an integrated decision-making framework as shown in Figure 4.2, which emerges from the continuous collaboration among individual, team, organizational, social, environmental, value co-creation, and situational awareness capacities.

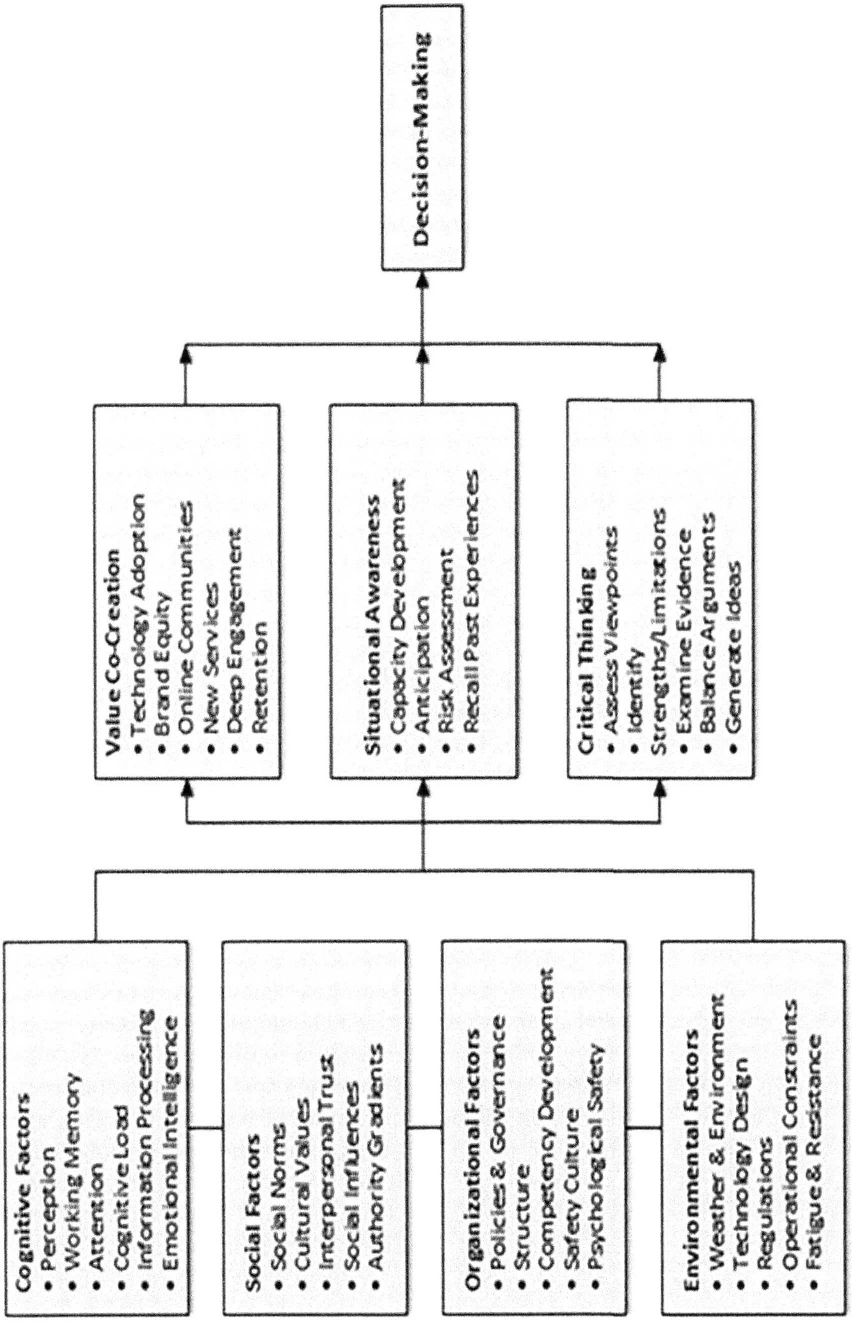

FIGURE 4.2 Integrated Decision-Making Framework.

4.4 INFORMATION GATHERING FOR DECISION-MAKING

Cognitive Systems Engineering (CSE) supports professionals to understand human factors in their decision-making processes. Information processing as a cognitive factor in decision-making refers to how individuals perceive, interpret, store, and retrieve information to understand a situation and choose an action (Shneiderman, 1988). Limitations in attention, memory, and processing capacity can lead to simplifications or biases, influencing judgement and increasing the likelihood of error under time pressure or uncertainty.

For example, engineers work for systems design to implement complex adaptive systems, and they need to take critical decisions in every step during the execution phase. It would be an effective way to consider relevant factors by observing cognitive and peripheral settings to identify alternatives and choose the best option to overcome uncertainties. This way, professionals can avoid errors or failure in executing projects or information system implementation.

Professionals need to gather information in the decision-making process when the task is critical, or the activities do not go as planned. Mistakes can occur as slips of actions or information gaps, which can cause disruptions in taking rational decisions.

Irrational decisions or improper procedures can cause adverse effects in systems implementation. Risks and errors can be anticipated and mitigated by identifying human factors. This is an initiative-taking human action that can allow professionals to maintain an implicit or explicit standard in making decisions with professionalism.

Professionals give more emphasis on gathering technical know-hows rather than acquiring adequate knowledge on human factors in different contextual situations, delivering rational decision-making in implementing complex adaptive systems. This exceptional knowledge enhances the productivity and skills of professional function as proactiveness to become judgemental decision-makers.

Gathering better information cues, addressing contextual experiences, identifying alternatives, giving feedback, and advancing analysis are applicable strategies for improved decision-making.

4.4.1 Cognitive Factor Analysis

Core cognitive factors for decision-making can be summarized as shown in Table 4.1, assessing perception, attention, memory, cognitive load, information processing, and emotional intelligence following different methodological approaches.

The assessment of factors influencing decision-making spans multiple domains – cognitive, social, organizational, situational awareness, value co-creation, and critical thinking – each requiring targeted methods to capture how individuals and groups process information, interact, and act in complex environments. For cognitive factors, methods focus on understanding internal mental processes such as perception, attention, memory, cognitive load, information processing, and emotional intelligence. Tools such as SAGAT and SART evaluate situational awareness and perception, while eye-tracking, dual-task paradigms, and NASA-TLX measure attention and cognitive load. Memory is assessed using recall/recognition tasks and scenario-based evaluations, and information processing is explored through think-aloud protocols,

TABLE 4.1
Summary of Information Gathering in Decision-Making

Factor	Characteristics	Definition	Methodology/Method
Cognitive	Perception	Sensory input (visual, auditory, tactile); Salience and attention-grabbing; Information cues; Misperception due to noise, ambiguity, or overload	Situational Awareness–based method; Eye-tracking analysis; Subjective rating scales; Think-aloud protocols
	Attention	Selective attention; Divided attention; Sustained attention	Eye-tracking; Situation Awareness Global Assessment Technique (SAGAT)
	Memory	To measure the encoding, storage, and retrieval of task-relevant information; Limited capacity ("7 ± 2" items, although realistically 3–4); Decays quickly under stress or overload; Experience-based patterns; Procedural memory (skills); Declarative knowledge (facts, concepts)	Recall and Recognition Tests; SAGAT; n-back or memory span tasks; Think-aloud protocols and task analysis
	Cognitive Load	High load → overload → errors; Low load → under-engagement → complacency	Subjective Rating Scales (NSA-TLX); Secondary Task Performance; Physiological Measures; Performance-Based Measures
	Information processing	To understand how individuals perceive, encode, integrate, and use information to make choices; Perceive, Interpret, Store, and Retrieve information	Think-Aloud Protocols; SAGAT; Eye-Tracking and Attention Metrics; Cognitive Task Analysis (CTA)
	Emotional Intelligence	To understand how individuals perceive, understand, regulate, and use emotions – both their own and others' – to guide effective decisions; Experience-based patterns Procedural memory (skills) Declarative knowledge (facts, concepts)	Mayer-Salovey-Caruso Emotional Intelligence Test (MSCEIT); Emotional Quotient Inventory (EQ-i 2.0); 360-Degree Feedback Assessments; Behavioural Observation & Scenario-Based Assessment; Physiological and Biometric Measures

(continued)

TABLE 4.1 (Continued)
Summary of Information Gathering in Decision-Making

Factor	Characteristics	Definition	Methodology/Method
Socio-Cultural	Social Norm	Shared rules or expectations within a group that guide behaviour; Reveals conscious and unconscious adherence to group rules; Teamwork and collaboration; Hierarchy and authority gradients; Social norms and group culture; Peer influence and groupthink; Trust, shared mental models, and communication quality	Questionnaires and Surveys; Vignette/Scenario-Based Experiments; Behavioural Observation; Implicit Association Tests (IAT)
	Cultural Value	Explains decision priorities shaped by cultural context; Deeply held beliefs and priorities shaped by culture, affecting how decisions are evaluated and choices are made	Surveys/Questionnaires; Interviews and Focus Groups; Ethnographic Observation
	Interpersonal Trust	Impacts collaboration, reliance, and information sharing	Trust Questionnaires/Scales; Economic/Behavioural Games; 360-Degree Feedback
	Social Influences	Shows how peer/group pressure shapes decisions	Social Network Analysis (SNA); Conformity/Peer-Influence Experiments; Surveys on Perceived Influence; Observation in Team or Group Settings
	Authority Gradients	The influence of hierarchical structures on communication, information sharing, and decision quality	Leadership and Hierarchy Perception Scales; Psychological Safety Surveys; Team-based Simulations; Structured Role-Playing Exercises
Organizational	Policy and Governance	Formal rules, procedures, and governance frameworks that guide decision-making in organizations.; Leadership style and governance policies, procedures, and rules Communication systems and reporting structures; Safety culture and psychological safety workload, staffing, and resource availability	Document Analysis/Policy Review; Surveys/Questionnaires; Interviews and Focus Group; Audit/Compliance Assessments

	Structure	The way work, roles, authority, and reporting lines are organized within the organization	Organizational Network Analysis (ONA); Surveys/Questionnaires; Observation/Workflow Analysis
	Competency and Training	The knowledge, skills, and experience of staff in making informed decisions	Skills Assessment/Knowledge Tests; Simulation or Scenario-Based Exercises; Training Records Review; Self-Assessment/360-Degree Feedback
	Safety Culture	Shared values, beliefs, and attitudes towards safety that influence organizational behaviour and decision-making	Safety Attitudes Questionnaire (SAQ); Hospital Survey on Patient Safety Culture (HSOPS); Observation and Audits; Focus Groups/Interviews
	Psychological Safety	The shared belief that individuals can speak up, share ideas, or report concerns without fear of negative consequences	Questionnaire (Edmondson's Team Psychological Safety Scale); Interviews and Focus Groups; Observation in Team Settings; Scenario-Based Exercises
Value Co-Creation	Technology Adoption	Identifies adoption barriers and enhances digital engagement	Technology Acceptance Model (TAM) or UTAUT; Usage Analytics; Interviews/Focus Groups
	Brand Equity	Builds trust and loyalty, supporting co-creation	Surveys/Questionnaires (Aaker's Brand Equity Model); Net Promoter Score (NPS); Social Media/Online Sentiment Analysis
	Online Communication	Improves information sharing and coordination	Content Analysis; Surveys/Questionnaires; Network Analysis (Map interaction flows)
	New Services	Ensures relevance and adoption of innovations	Surveys/Questionnaires; Usage Data/Analytics; Pilot Testing/Prototyping
	Deep Engagement	Drives active co-creation and informed decisions	Behavioural Analytics; Surveys/Questionnaires; Qualitative Methods (Interviews or diaries)
	Retention	Measures long-term participation and value sustainability	Usage Analytics/Longitudinal Data; Surveys/Questionnaires; Churn Analysis

(*continued*)

TABLE 4.1 (Continued)
Summary of Information Gathering in Decision-Making

Factor	Characteristics	Definition	Methodology/Method
Situational Awareness	Capacity Development	Awareness of knowledge gaps Ability to self-correct Reflection and error recognition. Builds foundational skills for accurate SA.	Training Evaluations/Competency Assessments; Simulation Exercises; Knowledge Tests/Quizzes
	Anticipation	Perception of elements in the environment; Comprehension (meaning); Projection of future status; Supports forecasting and proactive decisions	Situation Awareness Global Assessment Technique (SAGAT); Scenario-Based Exercises/Simulation; Think-Aloud Protocols
	Risk Assessment	Enhances hazard recognition and mitigation	Scenario-Based Risk Exercises; Checklists and Structured Risk Assessment Tools; Decision-Making Simulations
	Recall Past Experience	Improves pattern recognition and informed decision-making	Scenario Recall Tasks; Post-Event Debriefing/Reflection Exercises; Simulation-Based Performance Evaluation
Critical Thinking	Assess Viewpoints	The ability to recognize, articulate, and consider multiple perspectives in a decision scenario	Scenario-Based Exercises/Case Studie; Think-Aloud Protocols; Reflective Journals/Essays
	Identify Strength/Limitation	The ability to critically assess options, resources, or arguments by recognizing advantages and disadvantages	Decision-Matrix Exercises/SWOT Analysis; Case Study Analysis; Structured Peer Review
	Examine Evidence	The capacity to assess, interpret, and weigh relevant data, facts, or information for decision-making	Critical Appraisal Exercises; Simulation with Evidence-Based Tasks; Structured Problem-Solving Tests
	Balance Argument	The ability to integrate different perspectives, evidence, and risks to reach a reasoned judgment	Decision-Making Simulations/Role-Play; Structured Debate Exercise; Written Case Analysis
	Generate Ideas	The capacity to develop novel, feasible, or multiple solutions to problems	Brainstorming/Idea Generation Exercise; Problem-Solving Simulations; Divergent Thinking Tests

performance metrics, and cognitive task analysis. Emotional intelligence is typically measured through MSCEIT or EQ-i questionnaires, complemented by behavioural observations or scenario-based exercises.

For social and socio-cultural factors, assessment methods examine interpersonal and group dynamics, including social norms, cultural values, trust, social influence, and authority gradients. Surveys, scenario-based experiments, interviews, focus groups, and behavioural observation capture perceptions and behaviours in social contexts, while tools like social network analysis map influence and communication flows. Authority gradients and psychological safety are evaluated through structured simulations, observational studies, and validated questionnaires such as Edmondson's Team Psychological Safety Scale, highlighting how hierarchy, peer influence, and group norms impact decision-making. Organizational factors – including policies, governance, structure, competency, safety culture, and psychological safety – are assessed via document analysis, surveys, audits, simulations, and performance evaluations, revealing how systemic structures shape information flow, accountability, and risk perception.

Situational awareness, value co-creation, and critical thinking assessments focus on higher-order decision-making abilities. Components of situational awareness such as capacity development, anticipation, risk assessment, and recall of past experience are measured using training evaluations, simulations, SAGAT, and debriefing exercises, while value co-creation factors like technology adoption, brand equity, online communication, engagement, and retention are captured through usage analytics, surveys, social media analysis, and pilot testing. Critical thinking elements – including consideration of multiple viewpoints, evaluation of strengths and limitations, examination of evidence, balancing arguments, and idea generation – are assessed via scenario exercises, debates, think-aloud protocols, problem-solving simulations, and divergent thinking tests. Together, these methods provide a comprehensive, multi-dimensional approach to evaluating decision-making capabilities across individual, social, organizational, and technological contexts, offering practical insights for improving decision quality, reliability, and effectiveness.

How professionals perceive cognitive factors by detecting and recognizing information through sensory input (visual, auditory, tactile), salience and attention-grabbing cues, misperception due to noise, ambiguity, or overload. These characteristics determine what information gets noticed and what gets missed, directly affecting decision-making. Then, the ability to concentrate on relevant cues while filtering out interruptions through selective attention, divided attention, and sustained attention. The impact of these characteristics focuses on limited attention capacity that can lead to oversight errors or tunnel vision.

Working memory is another critical factor that explains short-term holding and tempering of valuable information demanded for the task. The memory with limited capacity ("7 ± 2" items, although realistically 3–4) can destruct professionals and decays quickly under stress or overload (Baddeley, 2012). These characteristics affect their ability to assess alternatives, track multi-level activities, stored knowledge, schemas, and rules, and recognize patterns and accurate intuitive decisions to maintain rational decision-making.

4.5 IMPLICATIONS AND INSIGHTS

Mental workload plays a critical role in human performance and decision quality, particularly when it exceeds or falls below an individual's cognitive capacity. Excessive workload can lead to cognitive overload, increasing the likelihood of errors by reducing reasoning capability, slowing information processing, and encouraging reliance on heuristics – mental shortcuts that expedite decisions but introduce systematic bias. Conversely, insufficient workload may result in under-engagement and complacency, diminishing vigilance and situational awareness. Cognitive biases such as confirmation bias, anchoring, availability bias, overconfidence, and framing effects often emerge under these conditions, shaping intuitive judgement without conscious awareness. These biases are intricately linked to core cognitive processes, including perception and attention, working and long-term memory, problem-solving, metacognition, and the effects of stress and cognitive load on situational awareness.

The implications for human factors are significant, as cognitive elements fundamentally determine how decision-makers interpret information, anticipate consequences, and select actions. An integrated human factors perspective recognizes that decisions do not occur in isolation but are shaped by the interaction of internal cognition, organizational design, social dynamics, and environmental conditions. This systems-based view is captured in established frameworks such as HFACS (Human Factors Analysis and Classification System), the SHELL model, SEIPS (Systems Engineering Initiative for Patient Safety), and Reason's Swiss Cheese Model, all of which demonstrate that decision errors typically arise from the alignment of multiple contextual factors rather than individual failure alone (Shappell & Wiegmann, 2000; Edwards, 1972); Carayon et al., 2006; Reason, 2000).Understanding these interactions provides critical insight into designing safer systems, improving decision support, and mitigating human error in complex and high-risk environments.

Social factors in decision-making arise from interactions among individuals and within groups, shaping how information is shared, interpreted, and acted upon. Elements such as team dynamics, shared mental models, organizational culture, interpersonal trust, communication quality, and hierarchy or authority gradients strongly influence collective judgement. Peer influence and social pressure can foster collaboration and coordination, but they may also increase the risk of conformity and groupthink, particularly in high-stakes or hierarchical settings. These dynamics directly affect whether individuals feel psychologically safe to question assumptions or voice concerns, making social factors central to the quality and robustness of decision outcomes.

The implications of social factors extend beyond interpersonal relationships to include the broader social and operational environment in which decisions are made. Effective teamwork and collaboration depend on clear communication, mutual trust, and aligned mental models, all of which enable individuals to challenge ideas constructively and arrive at shared decisions. At the same time, environmental and operational conditions – such as workspace design, noise, lighting, technology reliability, time pressure, fatigue, and physical stressors – interact with social dynamics to influence attention, situational awareness, and cognitive load. Poorly designed environments can exacerbate social pressures, reduce information exchange, and increase reliance on heuristic or rushed decisions. Together, these insights highlight the need for decision environments that support open communication, balanced authority, and ergonomically and operationally sound conditions to enhance both individual and collective decision-making.

Organizational factors define the systemic context within which decisions are made, encompassing structures, cultures, and processes that guide behaviour across all levels of an organization. Elements such as governance frameworks, policies and procedures, leadership style, communication and reporting channels, workload distribution, resource availability, and training systems collectively shape how information is generated, shared, and interpreted. Leadership approaches – whether supportive, hierarchical, authoritarian, or distributed – play a particularly influential role in setting expectations around accountability, risk tolerance, and openness. These organizational conditions directly affect situational awareness, perceptions of risk, and individuals' willingness to report concerns or challenge unsafe decisions, especially in high-pressure or regulated environments.

The implications for decision-making are especially significant in high-reliability industries, where organizational design is deliberately aligned to minimize error and enhance resilience. A strong safety culture and important levels of psychological safety enable staff to speak up, share critical information, and question assumptions without fear of reprisal. Effective communication systems and well-defined reporting structures ensure that relevant information reaches decision-makers in a timely and accurate manner. Additionally, adequate staffing, manageable workloads, and continuous competency development support sound judgement under uncertainty. Together, these organizational factors influence decision quality by shaping information flow, accountability, and the capacity to challenge and correct decisions before risks escalate into failures.

A structured cognitive process underpinning situational awareness involves the perception of environmental elements, the comprehension of their meaning, and the projection of their future states. From an organizational perspective, this process has critical implications for decision-making quality, as situational awareness enables individuals and teams to anticipate risks, select appropriate actions, and prevent unexpected outcomes. Organizational factors such as effective communication systems, clear procedures, supportive leadership, and adequate training play a decisive role in sustaining situational awareness by ensuring timely information flow and shared understanding. When these conditions are well designed, organizations enhance their collective ability to foresee emerging threats and respond proactively rather than reactively.

Critical thinking in decision-making is grounded in reasoning and problem-solving processes that rely on deliberate and analytical thought. Techniques such as inductive and deductive reasoning, risk assessment, option evaluation, and trade-off analysis enable decision-makers to navigate complex, high-stakes situations with greater clarity and structure. While these processes support robust and defensible decisions, they are inherently effortful and time-intensive, which can limit their use under time pressure or cognitive strain. The implication is that organizations and systems must deliberately create conditions – such as decision support tools, clear frameworks, and adequate time buffers – that allow analytical reasoning to be applied where it matters most.

Metacognition and emotional influences further shape the effectiveness of critical thinking, particularly in dynamic and uncertain environments. Metacognitive awareness, that is, thinking about one's own thinking, allows individuals to recognize knowledge gaps, monitor biases, self-correct errors, and reflect on decision quality, thereby strengthening judgement and adaptability. At the same time, emotional and affective states such as anxiety, fatigue, fear, or confidence interact with cognition, influencing attention, risk perception, and decision speed. These insights highlight the need to acknowledge emotions as integral to decision-making rather than as disruptions, and to design practices that promote reflection, emotional regulation, and self-awareness to enhance critical thinking and decision resilience.

4.6 SUMMARY AND DISCUSSION

This chapter has examined decision-making through the lens of human factors, emphasizing that effective judgement emerges from the interaction of cognitive, social, organizational, and environmental influences rather than from isolated rational reasoning. Cognitive factors such as perception, attention, memory, situational awareness, cognitive load, heuristics, biases, reasoning, metacognition, and emotional states explain how individuals interpret information, assess risk, and select actions under uncertainty. These elements illuminate why errors occur, particularly when mental workload exceeds capacity or when unconscious biases distort judgement, and they highlight the limits of purely rational or data-centric decision models.

Beyond individual cognition, the discussion has shown that social dynamics and organizational structures play a decisive role in shaping decision outcomes. Team interactions, shared mental models, hierarchy, trust, and communication quality influence whether information is exchanged openly or suppressed through conformity and groupthink. Organizational factors, including leadership style, governance frameworks, safety culture, workload, training, and reporting systems further determine how risks are perceived, how accountability is distributed, and whether individuals feel psychologically safe to challenge unsafe decisions. These factors collectively shape the context in which decisions are made and significantly affect decision quality and reliability.

Environmental and operational conditions add another critical layer to the decision-making process. Physical settings, technology design, time pressure, fatigue, and operational stressors influence attention, situational awareness, and cognitive load, often increasing reliance on heuristics and intuitive judgement. When environments are poorly designed or overly demanding, even well-trained individuals and teams are more likely to commit errors. Conversely, supportive environments and well-designed systems can enhance vigilance, reduce cognitive strain, and improve decision performance in high-risk and time-critical situations.

Taken together, this integrated view reinforces the systems perspective of human factors in decision-making, as reflected in established frameworks such as HFACS, SHELL, and SEIPS. These models underscore that decision errors rarely stem from individual failure alone but arise from the interaction between internal cognition, organizational design, social influence, and environmental conditions. Understanding these interdependencies provides valuable insight for designing safer systems, improving decision support, and fostering resilient decision-making practices in complex and high-hazard domains.

BIBLIOGRAPHY

Baddeley, A. (2012). Working memory: Theories, models, and controversies. *Annual Review of Psychology*, 63, 1–29. https://doi.org/10.1146/annurev-psych-120710-100422

Beaubien, J. (2000). [Rev. of culture at work in aviation and medicine: National, organizational, and professional influences]. *Personnel Psychology*, 53(1), 215–218. https://doi.org/10.1111/j.1744-6570.2000.tb00198.x

Butler, H. A. (2012). Halpern critical thinking assessment predicts real-world outcomes of critical thinking. *Applied Cognitive Psychology,* 26(5), 721–729. https://doi.org/10.1002/acp.2851

Carayon, P., Schoofs Hundt, A., Karsh, B. T., Gurses, A. P., Alvarado, C. J., Smith, M., & Flatley Brennan, P. (2006). Work system design for patient safety: The SEIPS model. *Quality and safety in health care,* 15(Suppl. 1), i50–i58. https://doi.org/10.1136/qshc.2005.015842

Edwards, E. (1972). Man and machine: Systems for safety. In *Proceedings of the British Airline Pilots Association Conference*. London.

Endsley, M. R. (1995). Toward a theory of situation awareness in dynamic systems. *Human Factors*, 37(1), 32–64. https://doi.org/10.1518/001872095779049543.

Flin, R., O'Connor, P., & Crichton, M. (2008). *Safety at the sharp end: A guide to non-technical skills*. CRC Press. https://doi.org/10.1201/9781315607467

Kahneman, D. (2011). *Thinking, fast and slow*. Farrar, Straus and Giroux.

Norman, D. A. (2013). *The design of everyday things* (Revised ed.). Basic Books.

Reason, J. (1990). *Human error*. Cambridge University Press.

Reason, J. (2000). Human error: Models and management. *BMJ,* 320(7237), 768–770. https://doi.org/10.1136/bmj.320.7237.768

Shappell, S. A., & Wiegmann, D. A. (2000). *The Human Factors Analysis and Classification System—HFACS*. U.S. Federal Aviation Administration.

Shneiderman, B. (1988). Information processing and human-machine interaction: An approach to cognitive engineering. *Information Processing & Management*, 24(1), 103. https://doi.org/10.1016/0306-4573(88)90080-5.

Wickens, C. D., Gordon, S. E., & Liu, Y. (2004). *An introduction to human factors engineering* (2nd ed.). Pearson Prentice Hall.

5 Critical Obstacles

Cognitive Bias and Heuristics in Decision-Making

5.1 INTRODUCTION

Decision-making is an inherently complex and often imperfect process, shaped by a web of cognitive biases, heuristics, and externalities. Decision-making becomes a decisive factor in determining the success or failure of project outcomes within engineering practices, particularly in the management and implementation of complex adaptive systems. At its core, it involves selecting a course of action based on available information, defined objectives, and operational constraints. Yet this process is frequently compromised by uncertainties, cognitive biases, and contextual influences that can lead to irrational judgements and unexpected repercussions. In highly uncertain and potentially dangerous engineering scenarios, the ability to make accurate and prudent judgements constitutes the basic ability of professional engineering practice.

Judgement and decision-making (JDM) is a critical process in both engineering and management field as it often requires making decisions under uncertainty and pressure to produce effective and responsible outcomes. While remaining ethically mindful when making decisions, decisions are often influenced by human factors, cognitive bias, and uncertainties in real-world cases, and some of them which we can keep in mind before making any decisions, such as availability, anchoring bias, overconfidence, tunnel vision, framing effect, sunk cost fallacy, primary effect, and groupthink.

Crisis decision-making is an emerging topic in systems engineering literature. When uncertainty, urgency, and incomplete information prevail, individuals and organizations become dependent on heuristics-mental shortcuts that decrease mental effort. There is a cognitive burden, but it can also cause systematic errors. Such prejudices can be quite harmless in normal life, however in emergencies, they may mean the difference between tragedy and safety.

Making decisions is essential to organizational behaviour because it shapes long-term performance, ethical standards, and strategies. But bias-free or logical approaches are rarely used for decision-making. Instead, they are impacted by heuristics and cognitive biases, which are mental shortcuts that can simplify difficult issues but also affect judgement. Heuristics can be helpful in situations where there is uncertainty. However, if they are not used properly, they can lead to incorrect logic, unethical blind spots, and suboptimal results. Therefore, promoting thoughtful

DOI: 10.1201/9781003685449-5

judgement and moral decision-making in modern businesses involves recognizing and reducing these biases.

In engineering management, judgemental decision-making is indispensable, especially across high-stakes domains such as aerospace, autonomous vehicles, robotic surgery, and intelligent logistics. Errors in decision-making within these contexts can escalate into operational inefficiencies, financial losses, or significant reputational damage. These challenges can intensify in high-risk or emergency situations, where professionals must absorb and interpret large volumes of information under severe time constraints. In such circumstances, decision-makers must rapidly assess risks, anticipate potential outcomes, and initiate strategic actions aimed at minimizing adverse effects.

However, cognitive biases and heuristic shortcuts frequently obscure rational assessment, giving rise to decisions that exacerbate rather than resolve emerging problems. Psychological factors such as overconfidence, confirmation bias, and the availability heuristic can quietly distort judgement, reducing the effectiveness of crisis management efforts. Understanding the role and mechanisms of these biases is therefore essential for developing more structured, evidence-informed approaches to risk mitigation and emergency response in engineering practices.

In contemporary engineering and industrial practices, professionals are increasingly confronted with complex and high-risk situations that demand rapid yet rational judgement. As these challenges grow in scale and intricacy, the study of judgement and decision-making becomes ever more critical. A deeper understanding of how decisions are formed enables professionals to navigate uncertainty more effectively and to select courses of action that safeguard both operational objectives and human well-being. Yet decision-making is far from a purely rational process; it is continually shaped and often destructed by cognitive biases and heuristic shortcuts. When left unrecognized or unmanaged, these mental tendencies can give rise to errors with catastrophic outcomes in engineering operations.

Every engineering operation relies fundamentally on robust decision-making processes to maintain stringent safety standards and achieve operational efficiency. At its essence, decision-making requires choosing among alternatives while assessing potential risks, anticipated benefits, and unknown variables. In high-hazard sectors such as chemical manufacturing or nuclear electricity generation, the consequences of poor judgement can be severe, ranging from fatal incidents and financial losses to environmental degradation and regulatory sanctions. Thus, the stakes of decision-making in engineering environments are especially high.

Across many complex engineering contexts, decisions are shaped by uncertainty, competing demands, and cognitive limitations. These factors can undermine clarity of judgement and precipitate actions that threaten organizational sustainability, public safety, and community well-being. For this reason, understanding and recognizing cognitive biases and heuristics are essential to unfold decision processes for preventing risk and upholding ethical engineering practices.

The ideal of purely rational decision-making is frequently constrained by the very architecture of human cognition. Biases can distort risk perception, foster unwarranted confidence, and prompt insufficient protective measures. In industrial settings where risks are elevated, rational and high-quality decision-making is indispensable, yet decisions are often made under pressure, with incomplete information and limited time. Rather than relying exclusively on systematic analysis, managers and engineers frequently turn to heuristics with mental shortcuts that simplify complex problems but can also introduce systematic errors.

Decision-makers operate within inherent limits of cognitive capacity, time, and information, leading them to satisfy rather than pursue optimal solutions. While such shortcuts may be harmless in everyday contexts, in complex and high-risk environments they may culminate in severe accidents and organizational failures. Recognizing these constraints and their effects is therefore vital for strengthening decision quality, enhancing resilience, and ensuring the safety and integrity of modern engineering systems.

5.2 COGNITIVE BIASES

Decision-making does not always arise from purely rational deliberation. Navigating uncertainty in a constructive way is essential to rational decision-making, particularly when professionals assess the probabilities of unknown or uncertain situations. Yet people frequently overlook critical information or neglect to weigh externalities appropriately, owing to the occurrence of cognitive biases. Cognitive biases are deviations in thought processes that systematically lead scientists to incorrect conclusions, impacting research outcomes even when the research is of excellent quality. These biases are systematic in nature, often yielding frequent errors in decision-making, particularly in critical situations marked by uncertainties or limited informational cues. When decisions are made in the absence of sufficient data, the mind tends to compensate by inventing assumptions, intuitively filling gaps in knowledge in ways that may fail to perceive reality.

Cognitive bias, therefore, refers to the mind's intrinsic tendency to mislead us with misinterpreted information that makes irrational decisions to perform an action. It is a compelling phenomenon because it demonstrates that even highly analytical and experienced professionals remain vulnerable to patterned errors in thinking. Recognizing cognitive biases involves acknowledging the ways in which our perceptions can be subtly distorted and understanding how to counteract these tendencies to improve decision in a qualitative manner.

In another context, heuristics are useful problem-solving tools for professionals, notably in tackling uncertainties and critical situations with information constraints, as these heuristics can support them to make judgemental decisions. Professionals choose to use several types of heuristics when the amount of information exceeds cognitive processing limitations, time is limited, or both. However, heuristic mechanisms can create obstacles to process rational thinking, making decisions with optimum performance that creates worse consequences rather than resolve critical problems. The performance of rational decision-making becomes controlled by cognitive biases and heuristic shortcuts that create defined incorrect judgements. Biases negatively affect our perception of risks, and they cause us to become more relaxed while leading us to take insufficient protective steps.

In engineering contexts, where precision, safety, and reliability are paramount, cognitive biases can exert a profound influence on outcomes. The specific biases that emerge vary according to the individual decision-maker, the situational context, and the nature of the task. Different cognitive biases manifest based on the individuals and the situation. Some common cognitive biases are confirmation bias, overconfidence bias, optimism bias, normalcy bias, attentional bias, authority bias, groupthink, and anchoring bias. The following paragraphs introduce several of the most prevalent cognitive biases, accompanied by illustrative scenarios that demonstrate how they manifest in engineering decision-making.

5.2.1 Confirmation Bias

Confirmation bias is a type of bias that influences decision-makers/professionals to only research and interpret the information that confirms pre-existing beliefs. Confirmation biases represent systematic errors in human judgement that arise when professionals rely on mental shortcuts, or heuristics, to interpret information. While these shortcuts can simplify complex decision environments, they often lead to irrational or biased conclusions, even when relevant and accurate facts are available. Professionals frequently approach situations with prior expectations shaped by experience, training, or organizational norms. These pre-existing beliefs can reinforce the perception that certain actions are unquestionably correct or necessary to avert worst-case outcomes. As a result, predetermined preferences may subtly yet powerfully shape decision-making through the influence of confirmation bias. Therefore, confirmation bias can occur when decision-makers interpret information in a selective manner in order to support prevailing views while ignoring conflicting evidence.

One of the most pervasive manifestations of this phenomenon is confirmation bias. This bias is the tendency to favour information that confirms pre-existing beliefs while discounting contradictory evidence. In striving to maintain the appearance of successful or stable operations, professionals may overlook or discount signals that contradict their expectations or preferred courses of action. Information that aligns with existing beliefs is readily accepted, while contextual, ambiguous, or conflicting evidence is dismissed or undervalued. In such circumstances, teams may unconsciously filter out early warning signs that indicate emerging problems, instead relying on familiar experiences, established references, or selectively interpreted case studies.

Confidence in favour of evidence that confirms pre-existing opinions, coupled with the shunning of information that contradicts those same opinions, often leads to biased decision-making and reinforces existing beliefs regardless of objective facts. When this is in place, decision-makers prefer to listen to evidence that supports the story of welfare integrity and dismiss mounting citizen and advocacy group complaints. Early audit data on error rates can downplay or re-characterize as implementation issues as opposed to structural problems. Through this selective perception, leaders can stand in a position to uphold their belief in the scheme, and back up the messages made to the masses that the task can work and is worth their support.

A more deliberate, context-sensitive review of available data can counteract these tendencies. By actively seeking disconfirming evidence and examining problems through multiple perspectives, professionals are better positioned to identify root causes at an early stage and to implement effective corrective measures. Such reflective practices enhance both decision quality and operational resilience, enabling timely interventions before minor issues escalate into significant failures. Professionals can selectively focus on signals suggesting that the system remained under control, while downplaying repeated warnings and abnormal events.

BOX 5.1 CASE STUDY ANALYSIS FOR BOEING 737 MAX

On October 29, 2018, the aviation world was shaken by the catastrophic crash of Lion Air Flight 610, a newly delivered Boeing 737 MAX 8, which plunged into the Java Sea and claimed the lives of all 189 passengers and crew on board. The shock had scarcely subsided when, only months later, Ethiopian Airlines Flight 302 – another 737 MAX 8 met a similar fate, killing 157 people. Together, these two tragedies exposed profound concerns surrounding the design, certification, and regulatory oversight of the Boeing 737 MAX program. Central to both accidents was a malfunction in the aircraft's Manoeuvring Characteristics Augmentation System (MCAS), a software feature intended to enhance flight handling, but which instead became a fatal point of failure.

Within Boeing, confirmation bias emerged as a dominant mode of thinking. Management repeatedly discounted or minimized information that challenged their conviction that the 737 MAX was safe to manufacture and operate. In the aftermath of the Lion Air crash, rather than undertaking a searching and impartial examination of MCAS and reconsidering its fundamental design, the company largely confined its investigations to explanations that aligned with its preferred narrative such as pilot error or maintenance deficiencies. Even as some Boeing engineers voiced serious concerns about the software's behaviour and potential risks, their warnings failed to gain sufficient traction. This environment reinforced a corporate culture in which dissenting perspectives were marginalized and critical scrutiny was undervalued, ultimately resulting in inadequate safety improvements and the continued production of an aircraft whose flaws had not been fully addressed.

In the above-mentioned case study, designers and engineers of Boeing downplayed the severity of complex design and safety aspects of Boeing 737 MAX early on, making decisions based on information that confirmed prevailing beliefs rather than latest information. This can be observed in professionals who put off taking drastic measures even when cautioned by safety regulations and government policies.

5.2.2 Overconfidence Bias

Overconfidence bias refers to unwarranted faith in one's own judgements or abilities. Overconfidence bias leads individuals to overestimate their knowledge, abilities, or control over outcomes. This bias can influence the organization's terrible approach to safety by making the decision-maker believe that their judgement is correct, while they do not examine and resolve the root issues since they effectively manage these problems in the past. Professionals may, at times, place excessive faith in the effectiveness of systems they have implemented, underestimating the risks that accompany them. This confidence can become a quiet hazard, discouraging the adoption of additional safety measures because existing approaches are assumed to be sufficient. Such assurance often breeds an illusion of control, leaving decision-makers overly convinced of the soundness of their judgements. The belief that current decisions alone will prevent failure reveals how overconfidence can narrow foresight and restrain proactive efforts to mitigate risk. By developing greater awareness of the potential weaknesses in their decision-making processes, professionals may be better positioned to take further steps that strengthen safety and resilience. Overconfidence and confirmation bias can lead people to favour quickly recalled hypotheses and ignore disconfirming evidence, even if it's more informative.

When professionals overemphasize their skills, expertise, and knowledge, they may be drawn towards decisions that are both risky and flawed, shaped by unconscious bias. This form of overconfidence can permeate all levels of an organization, from leadership and executives to frontline staff, particularly in situations marked by uncertainty. Rigorous risk assessment and careful scrutiny of assumptions can expose hidden flaws and guide professionals towards more balanced, rational choices. Yet, overconfident professionals often believe they can manage risk through sheer judgement, relying on overestimation and optimistic forecasting. Such approaches tend to sacrifice diversification and sound analysis, resulting in irrational decisions. Overconfidence can dominate the decision-making process, producing high-risk outcomes unsupported by robust evidence or safeguards during project implementation.

BOX 5.2 CASE STUDY ANALYSIS FOR JEFFREY MURDER CASE

Criminal law is founded on the promise of justice achieved through careful, evidence-based reasoning and a continual awareness of human fallibility. Yet this ideal is often tested by the realities of legal practice, where cognitive bias, incomplete information, and institutional pressures can quietly shape outcomes. When these influences go unrecognized, they may steer investigations and trials away from objective truth, allowing confidence to replace caution and assumptions to harden into conclusions.

The Jeffrey Gilham case, one of Australia's most contentious criminal justice stories, vividly illustrates these dangers. In 1993, Stephen and Helen Gilham were brutally stabbed to death in their Woronora home, where their son Christopher was also found dead. Jeffrey, the only surviving family member, maintained that he killed Christopher in self-defence after witnessing him murder their parents and set the house on fire. This account was initially accepted, leading to a 1995 manslaughter conviction for Christopher's death and no charges for the murders of Stephen and Helen. However, lingering doubts within the extended family and renewed investigative efforts eventually resulted in Jeffrey being charged with his parents' murders more than a decade later.

At the 2008 trial, the prosecution placed heavy reliance on expert forensic opinion, particularly the claim that similarities in the stab wounds across all three victims pointed to a single offender. Fire behaviour and circumstantial timelines were also used to challenge Jeffrey's version of events. The defence argued that the wound-pattern evidence lacked scientific validity and introduced new carbon monoxide testing, suggesting that all victims were alive when the fire was ignited – evidence more consistent with Jeffrey's account. Despite these challenges, the jury convicted Jeffrey, and he was sentenced to life imprisonment.

The appeal marked a decisive turning point. The credibility of the expert testimony was dismantled, and the toxicological evidence revealing significant carbon monoxide levels fundamentally undermined the prosecution's theory. In 2012, the New South Wales Court of Criminal Appeal quashed the convictions and acquitted Jeffrey Gilham, concluding that contaminated expert evidence and new forensic findings had transformed the case. Today, the Gilham case stands as a powerful lesson in the perils of tunnel vision and expert overconfidence, reinforcing the ethical imperative for critical scrutiny, scientific rigour, and humility in criminal justice decision-making.

The overconfidence bias is a situation in which a professional exhibit unjustified belief in their decisions or technological products. When implementing an added information system, professionals can show overconfidence in the correctness and effectiveness of the new system. The senior professionals presume that technology can cut down on the possibility of human error and enhance compliance and may fail to note how it complicates the employment habits of welfare recipients. The arguments of legality and technical constraints that exists internally can be played down in favour of the idea that the automation is the way forward.

For example, the housing market offers a familiar illustration. Long regarded as untouchable and among the safest investment opportunities, it often inspires unwarranted certainty. Banks, operating under the assumption that property values would rise indefinitely, can exploit this confidence. When critical information is ignored or insufficiently examined, such overconfidence can spiral into catastrophic consequences for the banking sector. This example underscores the necessity of questioning assumptions and critically evaluating information to prevent overconfidence from distorting judgement.

5.2.3 Optimism Bias

Optimism bias is the trend where the decision-maker believes that the likelihood of terrible things happening is low, while overestimating the probability of positive events happening. Optimism bias is a concept that things will work out for the best, often seen in project development. It can lead to egotism when we become familiar with something, causing us to make assumptions that are not based on specific data.

Professionals may likely underestimate the risks associated with the recent technology, assuming it would perform similarly without adverse effects.

People systematically underestimate their risk of encountering negative events. This is reinforced on job sites where accidents are rare, and the successful completion of hazardous tasks becomes normalized, downplaying real risk. Professionals may be convinced themselves that the risk would be low, even if there is no solid data to prove it. This tendency can lead to optimism bias: because previous events with successful implementation had not caused a disaster, they can assume the decision would be safe again. This bias reduces their sensitivity to the warnings and allow them to move forward without assessing risks.

Overall, the optimism bias can create a dangerous situation where technical warnings are compromised or ignored. Instead of making cautious decisions under uncertainty, professionals can convince themselves that it is safe to proceed. That is why incidents can happen not only because of technical failures but also because of optimism bias.

For example, in the Boeing 737 MAX incident, Boeing executives assumed that MCAS was exceptionally reliable and that pilots would handle any unexpected behaviour. They underestimated the potential for system failure and that optimism bias was one of the major reasons for the incident.

Since it is assumed that the arrangement will work effectively, professionals can overlook the dangers of having adequate preparation of dealing with critical design considerations and systems implementation. They can significantly minimize the chances of failure, assessing contextual limitations and constraints with the assumption that catastrophic situations are unlikely to reverse back to normal conditions. It can have to pay more values and sacrifice many lives for negligence and lack of attention.

For another example, policymakers can often be lax and believe that the financial institutions could manage their risks, underestimating the severity of any sort of failure. The team can overly be optimistic about the financial operation's success and underestimate the risks of failure. This bias leads them to neglect contingency planning, making them unprepared for unexpected complications. A more balanced outlook could prompt them to implement additional safety measures.

This was particularly evident when professionals assume that any technical feature would remain functional despite obvious signs of wear and degradation. Their bias leads to an overestimation of the systems' reliability and an underestimation of potential failure, fostering a false sense of security that can justify extending its use well beyond its safe operating life.

In another example, during the liquidity crisis of 2001, Swissair's leadership overestimated the probability of securing timely financing and underestimated the speed at which operations could be disrupted. Optimism bias contributed to a false sense of control, leading to delayed activation of emergency measures. This misplaced confidence diminished the airline's ability to negotiate effectively with creditors and suppliers before the grounding.

Decision-makers underestimated the risks associated with modifying equipment without thorough risk assessment. This bias causes individuals to believe they are less likely to experience negative events compared to others.

5.2.4 Anchoring Bias

Anchoring bias is the tendency to be influenced by the first information that the person knows/learns. This makes the decision-maker connect their understanding of the problem's severity to the outcome of past events. Anchoring bias is a situation where individuals outweigh the first piece of information in producing a decision. This bias occurs when one piece of information dominates decision-making, even when opposing evidence exists. In the project execution, a chance to save costs becomes the anchor. Every other concern, such as the environmental protection or the cost of public health, can be overshadowed by project cost. Even after people begin complaining about inverse effects, professionals may continue to cling to the initial anchor of "cost savings," delaying corrective measures.

BOX 5.3 DELAYED CRISIS RESPONSE OF SWISSAIR WITH OPTIMISM AND ANCHORING BIAS

During the 1990s, Swissair, Switzerland's former national carrier, pursued an ambitious "Hunter Strategy" and invested in a number of financially troubled carriers with little considerations. Despite this, the board chose to persist with the expansion plan rather than divest failing assets or revise its strategic approach. In 2001, Swissair faced a liquidity crisis that threatened its operational viability. The collapse of Swissair was not solely the result of flawed strategic choices but also of systematic cognitive biases that shaped those decisions. Examining critical decision-making moments reveals how specific cognitive biases undermined rational judgement and effective governance.

During the liquidity crisis of 2001, Swissair's leadership appeared to overestimate the probability of securing timely financing and underestimated the speed at which operations could be disrupted. Optimism bias contributed to a false sense of control, leading to delayed activation of emergency measures. This misplaced confidence diminished the airline's ability to negotiate effectively with creditors and suppliers before the grounding.

The airline's historic reputation as the "Flying Bank" became an anchor point for decision-makers, fostering the belief that Swissair's past financial resilience would translate into present crisis endurance. This bias skewed their risk assessment, causing them to interpret liquidity threats as temporary rather than existential. In reality, the company's overleveraged position left it ill-equipped to absorb sudden shocks.

In relation to the example scenario of bushfires, most residents acknowledge their evaluation of risk on past fire experiences. If they remained and defended their homes during smaller, less intense fires in the past and survived, they guess that they could again. This tendency leads them to misjudge the ferocity and speed of the fires, which can be caused by extended drought, severe heat, and high wind. These kinds of comparisons with the previous events lead to the illusionary feeling of safety and postponed evacuation till it is too late. In some towns, people did not evacuate until they could see flames, and this significantly added to the risk to them and the rescue teams that can be deployed to save their lives. Adding anchoring bias can cause a sizeable number of them to not act until safe escape routes have already been violated.

In another example scenario, the building's years of existence and prior weeks of normal activities operate as anchor. Decision-makers might not change their thinking from these outdated anchors when latest information, such as cracks or warnings, emerges. "We ran yesterday; we can run today," they may reassure themselves. This tendency reduces the importance of new threats. In another context, banks and financial institutions can anchor on the previously stable mortgage rates and undermine the likelihood of new borrowers failing to meet their repayments.

5.2.5 Attentional Bias

Under pressure or fatigue, professionals can focus on salient, immediate tasks such as completing the job at the priority of accomplishment with less attention to safety regulation or regulatory obligations. This can divert the focus from identifying the critical problems in executing tasks. Attentional bias is well documented in ergonomics and human factors research.

The major decision in attentional bias is to focus early on testing or assessing user experience before the design process. This choice can be influenced by attentional bias, as similar defects in previous designs that can trace back to develop information systems. The team's initial judgement depends on the perceived experiences, making them confident to address all factors before the system implementation in a single case. This can lead to focusing on another defect or delay and might not solve it on time, such as when information system implementation does not follow conditions.

5.2.6 Authority Bias

Engineers and middle managers may raise critical concerns, but senior executives' opinions can dominate, suppressing others' opinions. Particularly in the hierarchical organizational structure, seniors apply authoritative powers using top-down approach. This can give rise to authority bias in decision-making.

Authority bias refers to the tendency to follow instructions from authority figures, even when those instructions conflict with professionals' own judgement. This reflects when supervisors order workers to return to the factory and workers who initially resisted but eventually returned. Therefore, the authority bias amplifies the power imbalance, silencing the most vulnerable decision-making. Decisions are unduly swayed by the person with expertise, discouraging challenge or verification by less knowledgeable supervisors or workers. Research shows organizational hierarchies often stifle "upward communication" about risk, especially in high-stress, blue-collar environments.

For an example, the Boeing-FAA relationship was melded by authority bias, which had serious detrimental effects on the accident. The FAA assumed that Boeing's competence could not be questioned by an independent check since it relied on Boeing's conclusion and internal assessment. This deference to Boeing prevented impartial external observers from examining the MCAS system's shortcomings and postponed regulatory action in the wake of the terrible incidents. In a similar vein, airlines overestimated Boeing's promises on pilot training, which increased the risk to both pilots and passengers.

For example, even engineers have adequate expertise, but systems warnings can be ignored once managers feel pressure from senior leaders. The private meeting without engineers can also create a situation of authority bias, where managers try to reach consensus that match organization's expectations, rather than listening to dissent. As a result, the technical data can be ignored, and the final decision would be based on organizational loyalty instead of safety.

5.2.7 Normalcy Bias

Normalcy bias causes professionals to underestimate the likelihood or severity of a disaster by assuming that conditions will continue as normal. Normalcy bias is a process in which an organization starts to allow and gradually accept a lower safety standard of the working environment as a normal thing, since the company believes that there will be no serious incidents that can happen in the future. This bias describes the tendency to underestimate the likelihood of a disaster occurring and to assume that everything will continue as usual. For instance, building regulators show this bias when they do not impose the closure of the building, as they assume that the situation could stabilize and that normal operations could continue. This approach to the problem is particularly dangerous in high-risk environments because it creates inertia when it comes to decision-making. In this case, this bias can cause regulators to fail to take decisive action that could have saved more than a thousand lives. In another context, warehouse workers assume that the presence of smoke would not be an immediate threat, believing it to be a minor operational issue rather than a precursor to an explosion. This bias leads individuals to underestimate the possibility of disaster.

5.2.8 Recency Bias and Halo Effect

The recent demonstration successes may prevail over technology choices (recency bias), and the halo effect may cause teams to overvalue products of prestigious vendors regardless of comparative evidence. The result is lock-in to high-profile technologies at the detriment of portfolio assessment. For example, new research on hydrogen emphasizes how it is too easy to invest in new pilot projects, but still business owners are reluctant in adopting innovative technologies rather than relying on existing solutions for chemical industries.

5.2.9 Cognitive Tunnelling/Tunnel Vision Bias

Cognitive tunnelling occurs when a person becomes fixated on one idea, piece of information, or solution that they ignore other important cues or alternatives. This fixation makes it harder to step back, re-evaluate, or adapt when circumstances change. In the cockpit emergencies for airplane, the pilots can experience information overload and confusion that lead to cognitive tunnelling. As professionals are determined, they can pay attention to completing a job, even when there might be a high-hazard risk, and they concentrate on performing their jobs.

Professionals can focus too much on the stringent implementation of a project and to fulfil the commitment of getting accomplishment even without considering alternative resources to make judgemental decisions. This can lead them to ignore the safety protocols and proper regulations, which should be their biggest priority in an ideal scenario.

Tunnel vision limits attention to a single goal, like meeting deadlines, while other factors fade into the background. Through tunnel vision, the building is seen as a machine for conducting commands rather than as a real structure with boundaries.

Confirmation bias and cognitive tunnelling are related concepts but are not precisely the same. Confirmation bias involves filtering and giving more weight to evidence that supports a particular hypothesis. In contrast, cognitive tunnelling is often used in problem-solving situations. It refers to the tendency to get fixated on a specific solution and stick with it, even if it is not working.

5.2.10 Status Quo Bias

Status quo bias is a psychological tendency to maintain the current state of affairs, even if change could lead to better outcomes. With this bias, the status quo serves as a reference point, and any change is perceived as a loss. This bias leads professionals to be overly optimistic about the likelihood of accidents. They assume that because risks have not occurred in the past, they won't occur in the future. Consequently, management persisted in maintaining its production practices and continues to choose high-risk production arrangements. This bias creates a dangerous operating model that systematically prioritizes production over safety and perpetuate it, blinding to growing risks.

Closely related to status quo bias, the tendency is to prefer continuity and resist disruptive change, even when change is necessary. There was a tendency to preserve the current state of operations rather than initiate disruptive change, even in the face of escalating risks. Supervisors and even regulators demonstrate this bias by pushing to maintain existing systems operation. The assumption continues daily operations as usual, rather than addressing the risk involved. This bias often arises in stressful situations because people perceive the costs of disruption to be higher than the uncertain benefits of change.

The company may overlook safer but more complex process improvements in favour of simpler modifications that seem to align with past operational strategies. Professionals can choose existing systems because they believe it simplifies their tasks even though they are aware that it can violate safety regulations. However, this established routine can make them resistant to change as they fail to acknowledge the dangers involved in this practice.

For example, there is a preference for sticking with airplane that is familiar. The 737-platform felt like a "known quantity," even though the new engine placement introduces complex aerodynamic challenges. In another context, financial regulators often act slowly because of a herd mentality. They assumed the financial sector to be self-sustaining and that undue intervention would only create panic in the market. The fact that only three major rating agencies lack any major competition only helped their cause.

5.2.11 Normalization of Deviance

Normalization bias refers to the situation where decision-makers gradually accept and adapt to risky behaviours that were originally considered unacceptable. This prejudice is prominently manifested in the daily operational decisions of an organization. The organization's illegal operations, such as excessive storage of dangerous goods and mixed storage of incompatible chemicals, may initially be just minor rule adjustments. However, over time, these behaviours can gradually be accepted by the organization and become normal operations. For instance, the unsafe storage condition of chemicals like nitrocellulose (the loss of wetting agent) can be accepted by daily operations for one chemical industry and can be adopted as a malpractice in a timely manner. This normalization deviation leads to the long-term existence of safety hazards and eventually causes accidents.

Small signs of danger or critical warnings are observed, but since no disaster will happen, professionals believe the risk would be acceptable. This bias makes them underestimate the seriousness of early warnings in the operational systems. Instead of treating warnings, they reinterpret it as proof that the system is safe enough. Over time, this can create a false belief that launches could continue without change.

Over time, an unsafe practice can be normalized at any company because no incidents occurred. Employees simply believe that since nothing bad happened before it must be safe to carry on working which leads to a lowering of safety protocols within the organization.

Complementing overconfidence would be the normalization of deviance, where unethical behaviour becomes routine and acceptable within an organization. Over several years, engineering teams and leadership can become desensitized to the risks of manipulating emissions systems, treating non-compliance as a manageable deviation rather than a violation of environmental law. Together, this bias can create a cognitive environment where unethical conduct is framed as rational problem-solving.

Normalization of deviance shows how unsafe practices progressively become accepted as standard when repeated without negative consequences occurring immediately. Organization can regularly tolerate risk-laden practice, and these includes actions such as bypassing safety checks and operating with known equipment flaws, engaging normalization of deviance. This is normalized due to the fact that prior operations can lead to visible disasters. Consequently, this cultural drift can erode safety margins and in turn embed complacency within the organizational routines. As noted, such patterns often intensify in contexts where economic incentives tend to override precautionary norms. Similar emphasis can be made on how repeated tolerance for anomalies creates false confidence in hazardous processes. Eventually, the normalization of such unsafe practices can reduce intervention thresholds, leading to a catastrophic system failure.

5.2.12 Availability Bias

People may think catastrophic collapses are unlikely because they don't see them regularly. Regular cracks, repair, and small disturbances are recalled in their memory. Advice from experts may be overshadowed by the narrative "we've seen cracks before." For example, recent housing trends were a confidence factor for lending, ignoring historical downturns.

5.2.13 Groupthink

Groupthink arises when group pressures for consensus override critical evaluation, leading to poor decisions. It shows that people use heuristic principles to simplify complex tasks. Heuristics are "rules-of-thumb" that make decisions easier but can lead to errors. The decision-making process may be influenced by groupthink, where the desire for consensus leads to a lack of critical evaluation of decisions. This can result in limited independent analysis and fewer challenges to flawed assumptions. Encouraging open discussion could mitigate this effect.

Groupthink occurs when the elements of group cohesiveness or alignment to political forces suppresses individual thoughts. For example, in government departments and in ministerial offices, employees who inquire about the legality or fairness of latest information systems are sidelined. The bureaucratic culture is such that it rewards adherence to its priorities at the ministerial level instead of questioning and inquiry and encourages self-review. Such unified conformism restrains in exploring alternate views and makes sure that decision-making is in no way vulnerable. There are many use-case scenarios, where subsequent investigation by the integrity commission showed that the officials were privy to the deficiencies and simply felt unable to criticize the generalization, but this is a typical feature of groupthink.

Managers and engineers avoid conflict, striving for consensus even after some of the engineers express concerns over possible failure before implementing projects. Sometimes, employees may hesitate to raise alarmist concerns if they perceive others in the group who remain calm or dismissive of the danger. This is the opposite of what healthy discussions should be like. Instead of encouraging different perspectives and opinions. the senior management usually takes decisions to force consent over the project execution and end up in a fatal disagreement. Decision-making is more than just picking a course of action.

For example, in the case of MCAS, groupthink was encouraged by Boeing's culture. The senior management was able to successfully prioritize speed, cost, and market control above open discussions about safety issues because of fierce competition with Airbus. Because the culture discouraged such dissent and favoured broad consensus on financial targets, engineers who questioned the MCAS lacked avenues for effectively elevating their concerns. Erroneous assumptions remained unquestioned in a setting devoid of diverse perspectives and critical discourse, which led to their being dangerously lowered along with safety to appease commercial pressure.

A series of irrational decisions can be made in a project team, with dissenting views suppressed and crucial evaluations neglected. The executives can often try to argue optimistically about constraints despite the criticisms from the staff, risking associations with job termination. Even the executives who know about the involved risk can choose reference and risk-based approach to raise their concerns. Eventually, this bias leads to the drawbacks of the project implementation and success without fatal disruptions.

5.3 HEURISTICS

5.3.1 Representativeness Heuristic

The representative heuristic occurs when people judge a situation based on how much it resembles something known rather than its distinguishing characteristics. Representative heuristic can influence professionals' decision-making by leading them to believe that past occurrences can result in a catastrophic explosion, so future incidents would be insignificant. Instead of considering these events as warning signs, organizations can recognize those events as manageable and insignificant. Representativeness and context effect give more importance to stereotypes than statistical probabilities in decision-making.

The professional team usually relies on past experiences and assumes that the current situation is similar to previous successful operations. This heuristic leads to a failure to recognize unique risks. By assuming that prior success would repeat under similar conditions, they overlook critical differences that can influence the outcome. Such reliance on historical patterns without considering situational variations increases the likelihood of error.

In the case of national water supply system, professionals assume that the city's water supply is just like other systems that do not require prompt corrosion management. They omit necessary treatments since the situation appears to be ordinary. Unfortunately, the water is significantly more corrosive, resulting in lead leaching into residences. The abnormal liquid level can be treated as similar to previous "minor" fluctuations, leading to an underestimation of the seriousness of the current condition.

The heuristic of judging a situation based on its similarity to a prototypical example usually works well. However, it can lead to bias when a situation appears slightly different from the typical example, even though its pattern of cues is similar.

BOX 5.4 REPRESENTATIVE BIAS ON CLEARON INCIDENT

On the 8th of December 2020, an explosion occurred at the Optima Belle facility of Clearon Corporation (Clearon), fatally injuring one employee, injury two more, and causing $33.1 million in property damage. The incident was investigated by the Chemical Safety Board (CSB), who published a report in 2023. The process occurring during the incident was the removal of moisture from the raw chemical, sodium dichloroisocyanurate (NaDCC) dihydrate (commercial name CDB-56), to produce NaDCC (commercial name CDB-63).

Around 5:30 PM, Clearon representatives left the facility for the evening. This decision left the operators on site with no one immediately available with experience drying CDB-56. There were a number of indicators that there was a decomposition reaction currently taking place, including significant chlorine being produced by the drying process and the black specks present in the samples taken after the Clearon representatives had left. While the Clearon representatives were reachable by texts and calls, they were unable to provide information on the observations being made on sight.

The biases that go into this decision include representative heuristic, which is based on the belief that a previously observed pattern of events is occurring due to a certain level of similarity with the current events; this is the predominant bias in this decision. The representatives had been on site and observed system behaviour consistent with their expectations based on previous experience drying CDB-56. This gave them a false sense of security that things would continue to be under control and behave as expected.

5.3.2 Availability Heuristic

The availability heuristic describes how people judge the likelihood of events occurring based on how easily examples come to mind. Availability heuristic is where professionals make probability choices based upon experience. Since a collapse of this magnitude never occurred before, many owners, regulators, and supervisors underestimated its probability. The lack of memory of similar disasters can contribute to the belief that operations could continue as normal.

Availability heuristic significantly shapes professionals' decision-making. This means professionals can consider a previous case first and compare it with current circumstances. Team members usually recall a recent issue caused by situations and overestimate the likelihood that it can occur again. As a result, more time and resources can be allocated to verify and validate uncertain situations, even though it engages only a minor contributing factor.

Decisions are influenced by past experiences where rapid draining or diversion can be effective, leading operators to assume that the same measure would work in this case. The biased interpretation of the data can lead professionals to conclude that the evidence for risk is "inconclusive" rather than "persuasive and urgent." This can create a false sense of security and form the foundation for all subsequent decisions that night.

The availability heuristic refers to decision-makers overly relying on information that is easy to recall or obtain to make judgements. This bias interacts to create a set of poor decisions. Professionals can follow readily available information, such as past success rates, rather than conducting a thorough risk assessment, which can lead to the availability heuristic. This can lead them to underestimate the potential for failure, as their judgement can influence and easily recall examples of positive outcomes rather than a balanced evaluation of risks.

We can explain the availability heuristic following fire safety issue in a warehouse. During the fire emergency response on an accident, on-site fire warden or fire safety manager apply most common fire experience (such as that for ordinary warehouse fires) to extinguish the fire, rather than using special-type fire extinguishing plans like those for "hazardous chemical explosions." Of course, warehouse may not report in the first instance that the fire was caused by hazardous chemicals. Due to the lack of accurate information on the types and quantities of dangerous goods in the warehouse, the fire warden could only make decisions based on general knowledge and experience, seriously underestimating the actual risk level of the fire scene. It can cause a large number of unnecessary casualties and economic losses. If previous minor fires or smoke incidents were resolved without catastrophe, employees may rely on those experiences to assess the current risk, underestimating the potential severity.

Professionals rely on what is mentally immediately available: the successful history of the relevant experiences. They recall readily available information, which gives rise to the availability heuristic. It can create uneven consequences and may damage property or human lives.

The availability heuristic also gives rise to risks that can trigger public concern, with reputational adherence influenced by publicity. Countries witnessing early high-profile outbreaks, such as building collapse or chemical explosion, have higher levels of public anxiety, whereas those witnessing delayed outbreaks are more distrustful of the virulence of accidents.

The likelihood of individuals or organizations judging an event based on examples or perspectives that come to one's mind often leads to biased risk assessment. For example, at Lehman Brothers, they chose to make recent investment due to their recent returns from MBS, real estate markets, and CDOs instead of looking at the current market conditions. This perception of the profit returns from the market was due to the easy availability of this information. Despite the signs of a negative market indicator, the executives at Lehman chose to invest in the MBS and real estate assets, disregarding the above information, which also boosted their confidence that they had the capacity to turn the market situation around.

In another example, The Dreamworld Thunder River Rapids Ride disaster of 2016, in which four people were killed following a catastrophic ride malfunction, stands as one of the most significant workplace safety failures in Australian history. Investigations revealed that the tragedy was not the result of a single error but rather the outcome of systemic deficiencies in safety governance, organizational culture, and decision-making at multiple levels. This assessment analyzes the disaster through the lens of judgement and decision-making, focusing on how heuristics including availability heuristic distorted risk perception and led to flawed decisions.

On 25 October 2016, a tragic accident occurred at Dreamworld on the Gold Coast, Queensland, when the Thunder River Rapids Ride malfunctioned and killed four people. The incident shocked the nation and exposed deep organizational failures in safety governance, decision-making, and crisis management (as shown in Figure 5.1).

The availability heuristic further reinforced this complacency. Decision-makers assessed the likelihood of catastrophic failure based on easily recalled past experiences, where similar faults had not resulted in fatalities. Because no major accident had previously occurred, they assumed the risk of a fatal incident was negligible. The inquest noted that this reliance on past experience, rather than formal hazard analysis, was a critical weakness in Dreamworld's safety culture. This bias explains why operations continued despite mounting evidence of mechanical unreliability. Risk perception

FIGURE 5.1 The Dreamworld Thunder River Rapids Ride Disaster.

Source: Author-Generated.

was distorted by routine exposure to faults, resulting in a culture of acceptance and inaction.

The most enduring consequence was reputational damage. Once positioned as a premier Australian theme park, Dreamworld became synonymous with negligence and tragedy. Public sentiment turned hostile, fuelled by media coverage that highlighted systemic safety failures and the company's prioritization of profit over protection. The company's brand image remains tarnished even years later, with many Australians continuing to associate Dreamworld with the tragedy rather than with safe family entertainment.

5.3.3 Framing Effect

Framing bias is a cognitive bias that refers to how information is presented and can influence decisions. It was introduced by Kahneman and Tversky in 1984. The way information is presented can impact decision-making and assessments of risk. Prospect theory suggests that decisions framed as gain lead to risk aversion, while those framed as loss induce risk-taking. However, risk in this context refers more to uncertainty than the outcome. The framing effect is making choices based on frames rather than information content.

The framing effect occurs when decisions are influenced by how information is presented rather than by objective facts. Framing is a common cognitive bias as people tend to alter the perception of information mostly in a positive way. This tends to mask the challenge as it is overlooked. The way in which risks and statistics are presented can influence public acceptability of information systems. For example, presenting vaccine efficacy in preventing serious disease rather than breakthrough infections was effective in increasing public confidence in vaccination.

If a feature is framed as a small handling aid, people accept fewer checks than if it is framed as a system that can move the system without efforts. The words we use change the level of scrutiny we apply. The presentation of a problem affects the decision. When this problem presented as a production emergency (like every hour we are closed costs us money), the solutions focus on keeping the operations going, such as shifting workers, starting later, and reducing loads. If this problem presented as a safety emergency ("people could die if we get this wrong"), the only sensible option becomes to close and evacuate.

The framing effect is the impact that the way information is framed has on decision-making. In Black Summer, formal bushfire warnings were organized as three tiers, namely, Advice, Watch and Act, and Emergency Warning. Even though these categories were intended to provide direction to the actions of the people, majority of the residents misconstrued their content. As an example, Watch and Act was supposed to mean that a fire was nearby, and things could change fast, and people should be prepared to evacuate at a short notice. But to many it sounded as though there was yet time to do something before immediate action was needed. When under stress, individuals tend to pay attention to the least threatening meaning, especially when language is indirect or technical. This misconception delayed the responses of the households and caused delayed evacuations, which predisposed more people to entrapment. Moreover, in cases where the residents perceived that the warning was not as urgent as it turned out to be, the residents would not trust future alerts. Such loss of trust is a big long-term issue because it can diminish the adherence to future warnings during other fire seasons.

In another context, many novice investors and home borrowers fell into the framing effect, where risky financial products were framed as new financial tools rather than experimental products. It describes how the synthetic brochure can be prepared and marketed to be the most stable and risk-free with maximum return to the public without including the mass risk involved if it fails. Losses loom larger than gains; the way information is framed can shift preferences – e.g., framing the network choice as avoiding catastrophic stock-outs (loss frame) versus minimizing cost (gain frame) can change cutover design and resilience investments.

In the case of the quality control mechanism, when the quality control team presents positive reports, the different stakeholders believe that everything would run smoothly based on these reports. The parties involved in the project do not question any aspect or situations that is abnormal. This can lead to the team overlooking the boulder and the gap present in the wall. The quality control team do not present reports that highlight the problems encountered in a clear manner. This could have altered the way in which the parties involved made their decisions in this unfamiliar situation. The manner in which information, reports, and findings can impact decision-making of the responsible parties can lead to bias in decision-making rather than objectivity.

The manner in which the progress of the excavation and construction of the diaphragm is reported by the quality control team can result in wrong interpretation of the actual situation and its seriousness. The quality control team can fail to report the obstruction encountered in a tone that warrant immediate attention. This can lead to the responsible parties progressing with construction of the retaining wall without addressing the challenge. The problem can escalate from one system to the next and eventually result in disaster.

5.3.4 Sunk Cost Fallacy

Escalation of commitments, often known as the "sunk cost fallacy" occurs when prior investments improperly influence current decisions. Escalation of commitment occurs when decision-makers persist in a failing course of action due to sunk costs, reputational concerns, or institutional inertia. The sunk cost fallacy occurs when individuals or organizations stick with a given action because of the resources they have already committed to it, even though it is no longer the most appropriate action to take.

If previous efforts to increase production are unsuccessful, the team may feel compelled to proceed with further modifications, even if risks were evident. The sunk cost fallacy refers to circumstances whereby decision-makers persist with unperforming strategies based on prior investments. For instance, Volkswagen had invested heavily in diesel technology as a foundation for its entry into the U.S. marketplace. Rather than abandoning this practice when issues started developing, officials escalated their investment even further in defending and perpetuating their use of defeat devices.

When the decision-maker continues to invest in the failing project despite signs of negative outcomes, it often leads to the sunk cost fallacy. It often happens to the firm or institution due to the fear of admitting failure or reputational concerns. In the production-oriented industries, companies invest enormous amount to establish factories and machineries, where this bias plays a vital role in investing further and operating costs even when the market would be collapsing. This skewed perception may signal weakness to the professionals and reputational concerns in the market as well. This bias can lead to a series of bad decisions and to bankruptcy.

When weighing the enormous economic and reputational costs of grounding market reality, companies tend to maintain operations and minimize changes rather than prioritize minimizing social risk, due to the resources already invested in development and market commitments. The persistence of Robodebt (a sophisticated national) system remaining an issue in spite of its widespread criticism is an indication of escalation of commitment, which is a bias in which decision-makers use the ongoing failures to compensate previous efforts. The commitment to save billions by using Robodebt had been made publicly, and political leaders were at reputational risk by going back on their word. Every new complaint made the scheme not only a policy mechanism but also an icon of financial restraint. Political cost would have come with an admission of failure; thus, leaders redoubled and compounded the damage and entrenched the image of toughness on the issue in the face of welfare fraud. Escalation of commitment therefore firmly entrenched bad decision-making far beyond the initial stereotype.

In the case of firefighting crews working through Black Summer, this occasionally involved keeping up defences on properties or fire lines that were unlikely to be salvaged. The emotional commitment to a site – and the time, effort, and equipment already expended – made it more difficult to pull out, even when circumstances were such that success was improbable. Such perseverance meant that resources could not be used in other areas that were just as or more vulnerable. In other instances, there was more focus on visible or well-known places at the expense of remote communities which might have had greater needs. Although this commitment frequently was motivated by a wish to assist and protect, it lowered the overall effectiveness of firefighting and decreased the capacity to adapt to the adaptive circumstances.

In the case of missile or rocket launch, it gives a national pride and sophistication for the nation. Usually, national agencies invest a lot of time, money, and reputation in shuttle programs, and each delay creates more frustration. Leaders give commitment to showing success, even when risks are clear. The presence of launching ceremony also creates social and political pressure, which acts like a symbolic event that "has" to happen. These pressures can reduce the ability of managers to evaluate the situation with objectivity.

For example, the Chandrayaan-2 mission, launched on July 22, 2019, was India's most anticipated space exploration project, costing approximately $145 million USD. This mission included an orbiter for long-term lunar observation, the Vikram lander named after Dr. Vikram Sarabhai – father of India's space program, and the Pragyan rover for surface exploration.

On September 7, 2019, over 100 million Indians stayed awake, gathered around television screens in homes, schools, and public squares. Some kids wore cardboard prop helmets; families prepared special meals, and the nation held its collective breath. They were watching something only three nations had done before – a soft landing on the moon. At 1:50 AM, when Vikram lander was just 2.1 km from making history, the screen went silent. The green line showing the lander's path froze. In that moment, India's lunar dream crashed onto the Moon's surface at 680 km/h, leaving behind broken metals but showing how smart people can make unfortunate mistakes.

The trajectory was launched from Satish Dhawan Space Centre using the GSLV Mk III rocket (Figure 5.2), followed by Earth-bound manoeuvres to position the spacecraft for lunar transfer. The spacecraft made five additional orbital adjustments after entering lunar orbit on August 20, 2019, demonstrating ISRO's mastery of orbital mechanics. On September 6, 2019, the Vikram lander separated from the orbiter, beginning its descent towards the landing site near the lunar South Pole at 70.9°S, 22.8°E.

The landing sequence had two phases: the "rough braking" phase – reducing velocity from 1680 m/s to 146 m/s while descending from 30 km to 7.4 km altitude, and the "fine braking" phase for final descent. The rough braking phase was

FIGURE 5.2 Chandrayaan-2 Launch Failure.

Source: ISRO, 2019.

successfully executed with all parameters normal. But during the fine braking phase at 2.1 km altitude, the lander's trajectory deviated from its path. Around 1:53 AM IST, communication ceased when it was nearly 335 meters from the lunar surface. 70.8810°S, 22.7840°E was the crash point, with components scattered across the site. This was first identified by Chennai-based engineer Shanmuga Subramanian and later confirmed by NASA's Lunar Reconnaissance Orbiter.

This analysis explains the Chandrayaan-2 mission not as failure, but as a study in human psychology and decision-making under extreme pressure. The normal bias made planners expect routine conditions despite attempting something unprecedented. Previous experiences with ISRO in predicting Earth and lunar orbits did not prepare them for this polar landing's unique risks.

The sunk cost fallacy grew stronger as time passed. When late simulations showed potential software problems, the project had already spent $120 million and five years. Public expectations were out of control. Emails like "we've come too far to stop now" made the decision-makers feel trapped. This psychological pressure prevented objective risk reassessment, pushing the mission forward despite growing concerns.

The immediate technical impact of biased decisions was the mission's failure at the most critical stage. The lander's crash at 2.1 km altitude was caused by the limited real-world testing and overconfidence in simulations which caused the software to glitch during fine braking. This autonomous system did not allow any human intervention, leaving no chance of recovery. Millions of dollars' worth of hardware was destroyed and ended the chance to study the south pole, delaying the research of water ice and polar geology.

5.4 USE CASE – TACOMA NARROWS BRIDGE COLLAPSE (1940)

5.4.1 Background of the Use Case

The Tacoma Narrows Bridge in Washington State opened to traffic on July 1, 1940. At that time, it was the third-longest suspension bridge in the world. The bridge was designed to be modern and elegant, but the use of narrow girders made it very flexible and unstable. During construction and early use, workers noticed that the bridge swayed and vibrated even in light winds, earning it the nickname "Galloping Gertie." Despite these warning signs, engineers and decision-makers did not make major changes. They relied heavily on experience and intuition instead of testing how the bridge would behave under strong winds.

Engineers ignored signs of danger, people's warnings and thought nothing will happen which further led to collapse of bridge (Figure 5.3). Engineers, officials, and designers were overly optimistic and did not pay attention to various views and doubts, and also believed that bridge is safe. The collapse led to huge economic loss, reputational damage to the engineers, and a big loss of public trust. Wind tunnel testing and stronger safety checks became part of bridge design. This case shows that engineer's choices and the way of decision-making were responsible for the result. Money saving, sophisticated designs, and opposition to people's views, observations, and opinions contributed to the failure.

5.4.2 Root Causes of the Incident

The bridge was designed in modern way of structure, and it was built to be light and cheap compared to other bridges. Hence, engineers used narrow deck and less material which made the bridge flexible. The bridge used to move a lot in light winds so people started calling it "Galloping Gertie." However, on November 7, 1940 winds of around 40 miles per hour caused the bridge to twist and move than ever before. As a result, the bridge entered a state called aeroelastic flutter, in which wind energy keeps adding to its motion until it could no longer stay stable. Finally, within few hours bridge broke apart due to strong twisting and fell into the water.

FIGURE 5.3 Tacoma Narrows Bridge Collapse.

Source: Washington State, 1940.

The collapse was not only an engineering failure but human one. While building the bridge, cost-saving and less material usage were major priorities, which pushed leaders and officials to choose a cheaper design. On the other hand, the public and politicians wanted a modern-looking bridge, which created pressure to complete it quickly. Few engineers had warned about stability and design issues, but these concerns were ignored and biases influenced the decisions. Anchoring kept everyone focused on the main budget, and safety measures were considered too expensive. Confirmation bias made people focused only on the design and groupthink results in no consideration of majority view of team members.

5.4.3 Identification of Cognitive and Heuristic Factors

Cognitive biases such as overconfidence, confirmation bias, and groupthink affected decisions. Human factors, including poor communication, lack of accountability, and pressure to finish quickly and cheaply, also contributed. At the time, the culture prioritized prestige and cost-saving over safety, which further increased the risks.

The project team chose much slimmer plate-girder deck with solid side panels instead of deeper truss deck which was a safer option. It further led the bridge to be cheaper, lighter, and attractive. On the other hand, it reduced its stiffness and made it less stable to resist wind pressure. The choices could have cleared in the safe direction if team had chosen strong truss design which was stable and costly, aerodynamic features to improve stability which was requiring additional weight and little high budget. Instead, the team chose the wrong path to design it in order to save money and build a modern-looking bridge during the Great Depression. Engineers also felt confident due to recent success with design bridges and theories.

Decision-making influenced by human, cultural, and cognitive biases are as follows:

- As per authority bias people trusted famous engineers and past projects.
- Confirmation bias states that evidence in favour of modern design was highlighted while warnings were ignored.
- Overconfidence played a worst role as engineers believed that much wind testing is not required and design will perform well.
- Structural engineers did not have proper aerodynamic knowledge and lack of wind studies at the time of project.
- Engineers did not work on risk assessment processes and future dangers.
- Modern values of structures created sleek and slim structure to save costs.
- Due to such decisions, the bridge was built with poor torsional stiffness and solid side panels which further worsened reaction to wind.
- Harmful twisting motion which happened uncontrollably led to aeroelastic flutter.
- To ensure the success of this project, they should have done wind tunnel testing, setting minimum stability requirements and independent design checks.

In July, quickly after opening, the bridge started showing unusual movements in the wind, even at medium speed. It used to move upward and then downward often, so it was nicknamed "Galloping Gertie" by people. Despite such dangerous movements of the bridge, instead of closing it and conducting a deep investigation, officials did small fixes by adding tie-down cables, introducing speed limits, and attaching dampers. They kept bridge open due to economic and reputational pressure by testing small fixes and monitoring tools. They did not pay serious attention to safety risks.

Decision-making influenced by human, cultural, and cognitive biases are as follows:

- The swaying and movement of such a huge bridge were treated as normal conditions until its collapse.
- Instead of rethinking the situation they thought about money and kept observing.
- Optimism bias diverted their attention to small fixes which was wrong as they were thinking wind cannot be reason for bridge collapse.
- Human factors played a crucial role, as bridge was not well instrumented and roles and responsibilities were unclear.
- Cultural factors were highly responsible for the bridge collapse because officials and leaders wanted to avoid embarrassment and financial loses and believed in small fixes.
- Many decisions went wrong, as the dampers failed to control the twisting and instability of the bridge.
- They should have closed the bridge instantly, considering the safety factors, and ordered a deep investigation.

On the crisis day, the winds were at 35 and 45 mph speed and moved the bridge from vertical bouncing to dangerous twisting called as torsional oscillation. Authorities took serious action on last day by closing it to traffic and to stop movement by adding cables. Observers recorded the whole event, in which one car got stuck on the deck. The dog in that car did not survive. Unfortunately, the twisting became more severe, the bridge entered a state of flutter, and eventually the bridge collapsed due to wind energy. Officials had to decide the correct timing to close the bridge as well as the type of repairs required in fierce winds, along with how to communicate with the public. Stress and uncertainty shaped their judgement and action bias led them to repair and fix the problems even as chances of success were low. It made conservative decisions harder to justify due to lack of clear understanding of torsional flutter.

Decision-making influenced by human, cultural, and cognitive biases are as follows:

- Lack of rules to closing bridge, no correct monitoring tools, and danger faced by workers while trying to fix problems in extreme winds.
- The culture of keeping infrastructure running without considering actual cause or future danger and pressure to leadership to manage the situation.
- It was a clear engineering and knowledge failure, though fortunately no people were killed.
- This bridge failure changed engineering forever in an optimistic way by including wind tunnel tests, stricter safety rules, standard bridge design, and aerodynamic studies.
- The better decision would be strict and much more wind testing in high winds, no fixes during the crisis, and structured emergency procedures.

The above-mentioned cognitive biases and heuristics are illustrated in the table below (Table 5.1).

5.5 CONCLUSION

In conclusion, decision-making stands at the core of both engineering and management practice, particularly in environments shaped by uncertainty, time pressure, and high stakes. This chapter has explored how cognitive biases and heuristics subtly yet powerfully influence professional judgement, often steering decisions away from rational and evidence-based outcomes. By examining biases such as confirmation, overconfidence, optimism, anchoring, authority, normalcy, and groupthink alongside heuristics including representativeness, availability, framing effects, and the sunk cost fallacy, the discussion has illustrated how well-intentioned experts can overlook early warning signs and technical signals. The analysis of the Tacoma Narrows Bridge collapse demonstrates how entrenched assumptions and information filtering suppressed critical insights, leading to catastrophic consequences when deviations from safe practice become normalized.

Effective decision-making in engineering demands more than technical competence; it requires conscious awareness of the psychological and organizational

TABLE 5.1
Summary Table of Cognitive Biases and Heuristics

Cognitive Biases	Heuristic
Overconfidence Bias – Engineers were too sure that the bridge design would work, even though wind effects were not fully understood	Availability Heuristic – Decisions were based on past bridge projects that worked, while new risks were ignored
Confirmation Bias: The case states that evidence in favour of new design was highlighted while warnings were ignored	Experience-Based Heuristic – Engineers relied too much on past experience and not enough on new scientific studies
Sunk Cost Fallacy – Since a lot of money was already spent, officials were unwilling to rethink or delay the project	Simplification Heuristic – Complicated aerodynamic problems were treated as simple structural issues
Groupthink – Team members stayed quiet and did not question choices to keep peace and move the project forward	Familiarity Heuristic – Engineers chose old, familiar methods instead of trying new aerodynamic solutions
Optimism Bias – Engineers believed that extreme wind events were unlikely and that the bridge would stay safe	Recognition Heuristic – Well-known structural fixes were chosen over newer aerodynamic methods
Communication Bias – Poor sharing of information and weak coordination meant risks were noticed too late	Shortcut Heuristic – Faster and cheaper design decisions were made instead of doing proper testing
Authority Bias – People trusted the lead experts without asking enough questions or challenging their views	Availability of Past Success – Because many other bridges had survived, engineers assumed this one would too
Status Quo Bias – People preferred to continue with the chosen design rather than explore safer but untested options	Representativeness Heuristic – The bridge was assumed to act like earlier suspension bridges, even though it was thinner and more flexible
Confirmation Bias – Early warning signs, like vibrations, were dismissed because they did not fit the belief that the design was safe	Similarity Heuristic – The bridge was treated as the same as older bridges, even though it was very different
Hindsight Bias (after the collapse) – After the failure, many said the problem was obvious, forgetting that knowledge was limited at the time	Anchoring Heuristic – Once the first design decisions were made, later evidence was ignored or undervalued
Escalation of Commitment – Even when problems appeared, leaders kept pushing forward to avoid delays and embarrassment	Rule-of-Thumb Heuristic – Standard formulas and practices were used instead of adjusting to the unique design

forces that shape judgement. Decision-makers must continuously negotiate technical constraints, stakeholder expectations, and safety imperatives while confronting invisible cognitive and institutional pressures. Recognizing that biases are not merely individual flaws but are often reinforced within hierarchical and cultural systems provides a foundation for developing mitigation strategies and resilient decision frameworks. By fostering reflective practices and bias-aware organizational cultures, professionals can ensure that risks are acknowledged, warnings are heeded, and responsible decisions prevail even under extreme pressure.

BIBLIOGRAPHY

ABC (Australian Broadcasting Corporation). (2016, October 25). Dreamworld: Four people killed on Thunder River Rapids ride at Gold Coast theme park. *ABC News*. www.abc.net.au/news/2016-10-25/four-people-dead-ride-at-dreamworld-gold-coast/7963880.

ABC (Australian Broadcasting Corporation). (2018, June 25). Dreamworld tragedy coronial inquest timeline. *ABC News*. www.abc.net.au/news/2018-06-25/dreamworld-tragedy-coronial-inquest-timeline/9900552.

ABC News. (2011, November 28). Murder appeal begins over Sydney family slaying. www.abc.net.au/news/2011-11-28/murder-appeal-begins-over-sydneyfamily-.

ABC News. (December 1, 2011b). Convicted murderer Gilham may get retrial. www.abc.net.au/news/2011-12-01/convicted-murderer-gilham-may-getretrial/.

ABC News. (December 1, 2011c). Convicted murderer Gilham may get retrial. www.abc.net.au/news/2011-12-01/convicted-murderer-gilham-may-getretrial/.

ABC News. (December 2, 2011d). Gilham released pending murder retrial. www.abc.net.au/news/2011-12-02/gilham-appeal-succeeds/3708630.

APS News. (2016, November 1). November 7, 1940: Collapse of the Tacoma Narrows Bridge. *APS News*. American Physical Society website American Physical Society.

Arioli, G., & Gazzola, F. (2015). Torsional instability in suspension bridges: The Tacoma Narrows Bridge case. *Communications in Nonlinear Science and Numerical Simulation*, 42, 342–357 https://doi.org/10.48550/arXiv.1508.03200.

Bagla, P. (2018). Space exploration India plans to land near moon's south pole. *Science (American Association for the Advancement of Science)*, 359(6375), 503–504. https://doi.org/10.1126/science.359.6375.503.

Baik, S. (2015). *The Wiley-Blackwell handbook of judgment and decision making* (G. Keren & G. Wu, Eds.; 1st ed.). Wiley Blackwell.

Bazerman, M. H., & Moore, D. A. (2013). Judgment in managerial decision making (8th Board Matters. (2020). Dreamworld nightmare: The coroner's report. Board Matters. https://boardmatters.com.au/insights/dreamworld-nightmare-the-coroners-report.

Board of Investigation. (1941). *Tacoma narrows bridge: Report of the board of investigation (Part 1 & 2)*. Washington State Department of Transportation.

Chen, J. (2024, May 2). *Heuristics: Definition, pros & cons, and examples*. Investopedia.

Clayton Utz. (2020, August). The Dreamworld tragedy: The Coroner's findings, the prosecution and lessons learnt. *Clayton Utz Insights*. www.claytonutz.com/insights/2020/august/the-dreamworld-tragedy-the-coroners-findings-the-prosecution-and-lessons-learnt.

CSB. (2023, July 7). CSB Releases Investigation Report Into Fatal 2020 Explosion at Optima Belle Facility in Belle, West Virgina. *Targeted News Service*. www.proquest.com/docview/2834081800?accountid=17095&parentSessionId=86rAzdrOaV8ibD6DpLakTpygDFnWrbtFaMykzfm32jg%3D&pq-origsite=primo&searchKeywords=Fatal%20chemical%20decomposition%20reaction%20and%20explosion%20at%20optima%20belle&sourcetype=Wire%20Feeds

Davies, P. B. L. H. L. (2012, June 25). *The verdict after 19 years: Gilham killed one decisions and fiascoes*. Houghton Mi lin. dihydrate#section=Information-Sources.

Disaster Breakdown. (March 20, 2024). The Boeing 737 Max crashes: A comprehensive analysis – disaster breakdown. YouTube. www.youtube.com/watch?v=lDGObEGgOXE.

Dragomir, V. D. (2021). Political interventions in state-owned enterprises: The corporate governance failures of a European airline. *Journal of Accounting and Public Policy*, 5, Article 106855. https://doi.org/10.1016/j.jaccpubpol.2021.106855.

Ellis, D. (2023). Parallels between Swissair and Etihad Airways. *Journal of Air Transport Management*, 45, Article 102000.

Gazzola, F., Jleli, M., & Samet, B. (2022). A new detailed explanation of the Tacoma collapse and some optimization problems to improve the stability of suspension bridges. *AIMS Mathematics and Engineering*, 10(7), Article 3045. https://doi.org/10.3934/mine.2023045.

Green, D., & Unruh, W. G. (2004). Tacoma bridge failure—A physical model. https://doi.org/10.48550/arXiv.physics/0408101.

Guptan, M. (2019). How did Chandrayaan 2 fail? ISRO finally has the answer. *The Week*. www.theweek.in/news/scitech/2019/11/16/how-did-chandrayaan-2-fail-isro-answer.html. Accessed Dec 2019.

Guyer, J. P. (2010). *Ethical issues from the Tacoma Narrows Bridge collapse*. CED Engineering. www.cedengineering.com.

Indian Space Research Organisation. (2019). *Chandrayaan-2 mission overview*. Department of Space, Government of India.

Janis, I. L. (1972). Victims of groupthink: A psychological study of foreign-policy decisions.

Kahneman, D., & Tversky, A. (1979). Prospect theory: An analysis of decision under risk. *Econometrica*, *47*(2), 263–291. https://doi.org/10.2307/1914185

Kim, S., Kim, K. J., & Yi, Y. (2022). Investigation on Lunar landing candidate sites for a future lunar exploration mission. *International Journal of Aeronautical and Space Sciences*, 23(1), 221–232. https://doi.org/10.1007/s42405-021-00433-4.

Kumar, A., & Raj, R. (2020). Analysis of Chandrayaan-2 mission: Technical challenges and organizational learning. *Journal of Spacecraft and Rockets*, 57(4), 789–802.

Medina, C. A., Cannon-Patron, S. D., Thompson III, S., Haneen, F., & Gourdet, S. (2024). A case study of management shortcomings: Lessons from the B737-Max aviation accidents. Student Research Symposium (SRS), Embry-Riddle Aeronautical University. https://commons.erau.edu/db-srs/2023/poster-session-two/6/.

Moore, D. A., & Healy, P. J. (2008). The trouble with overconfidence. *Psychological Review*, 115(2), 502–517. https://doi.org/10.1037/0033-295X.115.2.502.

Samuelson, W., & Zeckhauser, R. (1988). Status quo bias in decision making. *Journal of Risk and Uncertainty*, 1(1), 7–59. https://doi.org/10.1007/BF00055564.

Staw, B. M. (1981). The escalation of commitment to a course of action. *The Academy of Management Review*, 6(4), 577–587. https://doi.org/10.5465/amr.1981.4285694.

Sterman, J., & Quinn, J. (July 26, 2023). Boeing's 737 MAX 8 disasters | MIT sloan. mitsloan.mit.edu. https://mitsloan.mit.edu/teaching-resources-library/boeings-737-max-8-disasters.

Thompson, D. (2024). Boeing 737 max crisis of 2018-2019: A content analysis. In *Proceedings of the 37th International RAIS Conference on Social Sciences and Humanities, vol. 2* (pp. 55–70). Scientia Moralitas Research Institute https://doi.org/10.5281/zenodo.13916413.

Tufano, A. V. (1960). *Investigation of wind forces on highway bridges*. ProQuest Dissertations & Theses.

Tversky, A., & Kahneman, D. (1973). Availability: A heuristic for judging frequency and probability. *Cognitive Psychology*, 5(2), 207–232. https://doi.org/10.1016/0010-0285(73)90033-9

Tversky, A., & Kahneman, D. (1974). Judgment under uncertainty: Heuristics and biases. *Science*, 185(4157), 1124–1131 https://doi.org/10.1126/science.185.4157.1124.

Tversky, A., & Kahneman, D. (1981). The framing of decisions and the psychology of choice. *Science*, 211(4481), 453–458. https://doi.org/10.1126/science.7455683.

6 Uncertainties for Human–Computer Interaction (Heterogeneous Environment) in Decision-Making

6.1 INTRODUCTION

Professionals are requiring access to a wide spectrum of various digital applications/services without being limited by constraints such as time, location, technology, device, and mobility restrictions. Recent research in the area of user experience has shown that such user experience (UX) mechanisms may need to be complemented with more user-centric approaches in order to truly meet end-user engagement and expectations for the interactive services, particularly in heterogeneous environment. Real-time UX prediction and dynamic service retention for both human and machine in human–computer interaction (HCI) would be an essential factor to maintain next generation decision-making. The UX dataset involving the time and location constituted to make data-driven decisions to analyze performance and provide an optimal solution based on the insights derived from the UX.

It is important to note that, UX was at the bottom of priorities of the majority of service providers or manufacturers in different industries, as they gave more emphasis to expand network and enhance production performance to deliver optimum level of predefined services. Today's customers are becoming a different shape that demands customized solutions in the age of human and machine/things interactions through haptic communications. They require not only more quantity, but also a proactive approach from providers to self-manage their services from end-user terminals. For example, 5G accommodates an opportunity to add trillions of connected devices that enable smart applications such as smart cities, smart retail, intelligent transport system, smart logistics, IoT applications, etc., and it creates an inclusive environment for professionals to make proactive decisions to cater hyper-interactive services to end-users. 5G transmits much more data over wireless networks, which in turn can open the door to accommodate immersive and interactive digital services in various aspects. This hyperactive interaction provides the opportunity to analyze the decision-making process, considering users' behaviour and proactive measures.

DOI: 10.1201/9781003685449-6

Today's decision-making in HCI requires a combination of individualized insights, connected interactions and an agile approach to identify alternatives in critical situations. This means more than simply doing the same things over in the new situation with uncertainty. It requires new ways of exploring user trends and preferences and being smarter about responding to these factors before making decisions. The decision-making would be more challenging for user-centric services like smart city applications, smart shopping, IoT applications, autonomous driving, smart logistic services, fintech, smart home appliances, etc. So far, research on quality of experience (QoE) has mostly been conducted in the area of communications, and UX has addressed hedonic and pragmatic usage aspects of interactive applications. However, a further approach for a comprehensive view of analyzing decision-making by addressing pragmatic, hedonic, and eudaimonic aspects, such as meaningfulness and purpose of use, and by introducing the multidimensional construct of expectations is critical to organizational decision-making.

UX is defined as a momentary, principally evaluative feeling (good-bad) while dealing with a product or service (Hassenzahl et al., 2010). People perceive interactive product or service along with two different dimensions. Pragmatic value refers to the product's perceived ability to support the achievement of "do-goals." Pragmatic quality concentrates on its utility and usability in relation to potential tasks. On the other hand, hedonic value refers to the product's perceived ability to support the achievement of "be goals," such as "being competent," "being related to others," and "being special." Hedonic quality focuses on the self, i.e., the question of why someone owns and uses a particular product. In the context of technological system use, an application of this concept for exploring the positive experience of interactive products and services has found a clear relationship between need fulfilment and positive affect towards the (technological) systems. These eudaimonic aspects relate to something that is worth pursuing in life or something that is intrinsically worthwhile to human beings and are not automatically tied to subjective experiential states. Professionals need to examine pragmatic, hedonic, and eudaimonic experiences through HCI in the next generation decision-making.

Professionals can extend their analytical strengths by endorsing UX approaches so as to be able to assess the subjective analysis, considering HCI-based applications that support in making rational decisions. Existing models for UX are extended by adding the eudaimonic dimension as a factor that influences UX, and we incorporate user needs and motivation as factors that have an effect on the individual dimension aspects in the decision-making process.

Organizations are adopting smart interactive technologies that allow them to move forward for getting optimum level of performance, but there is a challenge for professionals in making rational decision in the smart landscape, where both providers and consumers use smart technologies to revamp and boost their roles in the service economy and eventually enhance their experiences. The HCI enhances organizations' capabilities with smart experience and new digital opportunities that also create uncertainties in critical situations. This smart heterogeneous environment is considered as "an emergence of a consequence as a vision, using modern technologies to add value for experience to improve human life with respect." The key features include the availability of products through advanced technologies, allowing customers to interact in scenarios, while tracking consumers' preferences and trends. In the smart environment, professionals need to address a versatile selection process for services with more options as well as customer preferences before making decisions.

6.2 HUMAN–COMPUTER INTERACTION

HCI engages interactive capabilities and enhances information systems' ability to get connectivity between human and non-human system. The HCI includes the convergence of advance technologies like sensors, AI, and big data to augment human life and expand knowledge that are the main catalyst to make it happen for the smart interaction between human and machine (Wobbrock & Kientz, 2016). Organizations are adopting this new trend for their business continuity, and it emphasizes on customer engagement, sentiment, and value co-creation in the service ecosystem. So, professionals can assess co-created values through HCI in their decision-making journey. 'Value co-creation" is termed as a process to gain value from customers' deep engagement, and users equally gain satisfaction through unique experiences in HCI. The HCI engages a collaborative co-experience and user interaction in the heterogeneous platform, where users interact with the system with multiple activities through multi-modal interactions. For example, virtual reality (VR) allows users to perceive sense of presence, also known as physical presence, to reflect and create a user's sense of "being there" or "being in a virtual place" to make user believe or feel that they exist within the virtual world. If professionals make any decisions regarding VR-based platforms, they need to assess sense of presence from user end and also key parameter indicators (KPIs) from the platform.

Nowadays, heterogeneous environment leads human-to-machine or machine-to-machine communication through HCI/Digital Twins. UX for digital services is reshaped through both the services and contextual scenario cases. Hyper-realistic digital services require high bandwidth, high data rate, and low latency to deliver without compromising service interruptions. Big data analytics (DA), artificial intelligence, cloud computing, Metaverse, mixed reality (MR), IoT, etc. play an enormous role in engaging HCI. Professionals need to analyze human factors like key quality indicators (KQIs) for the perceived usage and also quality of decision (QoD) for human–machine interactions as well as other hedonic and eudaimonic factors in their decision-making to get an optimum outcome.

With technological advancement, the interaction paradigm transforms its nature with a heterogeneous environment to get value from multi-modal interactions. Interactive features enable more sources of information to enhance the UX. So, the interaction paradigm for the heterogeneous environment enhances an interactive and responsive UX through multi-modal interactions. Digital platforms support intrinsic motivation to provide optimum outcomes and experiences to end-users.

The exponential growth of mobile data traffic, rapid increase in connected devices, visualizing large amounts of data, and hyper-realistic media are the primary enablers to enhance heterogeneous environments through an intelligent digital platform, and it demands a comprehensive model to assess the decision-making in such a hypersensitive UX platform through HCI.

The HCI is revolutionizing and becoming sustainable by incorporating smart systems and modern technologies to complement the whole transformation. Advanced technologies provide implications of HCI and its impact on the perceived quality of UX creates an enabling environment in the decision-making process. Professionals can assess QoE from the human–machine perspectives, particularly in the heterogeneous environment for making rational and judgemental decision-making. Even though organizations are integrating advanced technologies in information systems, professionals can execute decision-making models to identify alternatives before making decisions.

In the current computing models, such as fog, edge, and IoT, new digital services are related to the perceived end-user experience, and professionals can assess valuable insights for HCIs. Therefore, a comprehensive decision-making framework is to be measured for the performance evaluation in the heterogeneous environment. For example, network engineers can deploy 5G wireless network to get optimum value propositions that have many potentials like higher bandwidth, higher speed, low latency, faster data rate, and a low-cost solution resulting in the user perceived rate up to 1 Gbps with 1 ms round-trip latency as well as to reduce 90% of the network energy consumption.

How is the HCI paradigm formed? Industrial applications are accommodating emerging technologies in their operational activities to allow professionals through multi-modal interactions. Artificial intelligence (AI), augmented reality (AR), VR, Internet of Things (IoT), cloud computing, big data, 5G, etc. are emerging technologies that transform a paradigm shift to enhance organizational capabilities and it demands to adopt new decision-making models for professionals in making rational decisions. The process flow from system integration to KQIs in HCI is illustrated as shown in Figure 6.1.

The introduction of emerging technologies revolutionizes the transformation of a paradigm shift to enhance interaction capabilities for information systems and provide enormous opportunities through sensory information. However, these interactions can create uncertainty in critical situations. Unlike traditional service deliberations in critical environments, more specifically, where professionals face new contextual cases to resolve problems, they can apply established tools in the decision-making process.

For example, professionals can face critical situations in making decisions to optimize services using AI-driven recommendations for digital applications. AI co-creates values to render virtual realism for product visualization, interactive and predictive recommendations, responsive service optimization, and value extraction through online reviews that creates critical situations to make decisions, and professionals need to develop their knowledge of how to analyze AI-driven tools before making decisions. In the digital platform, AI engages with those virtual interactions and explores users' requirements to enhance customers' decision-making, and it equally creates an environment for professionals to show their potential in rational decision-making process. This requirement has been amplified due to rapid adoption of advanced technologies in customer expectations for heterogeneous HCI settings. In the age of AI, digital platforms use predictive DA, reflecting consumer preferences, network optimization, product visualization, and interactive recommendations to deliver optimum outcomes.

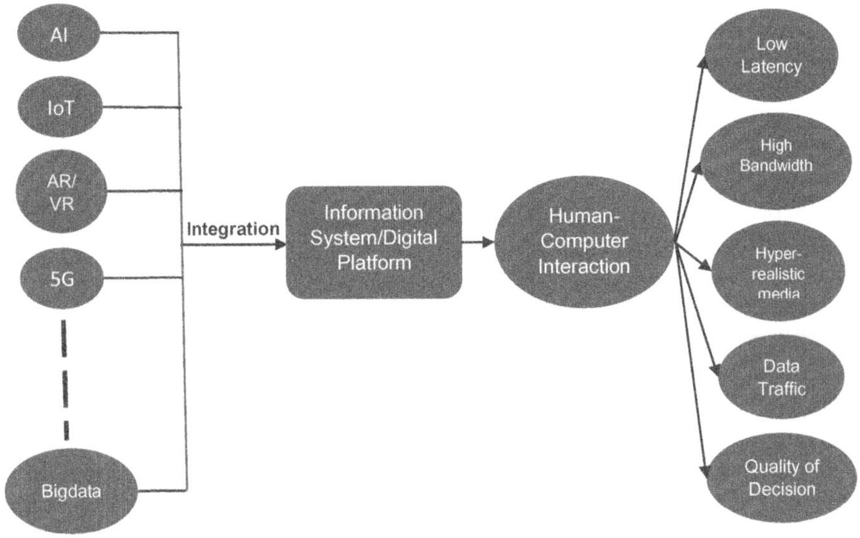

FIGURE 6.1 Process Flow Map for Human–Computer Interaction.

6.3 APPLICATIONS OF HUMAN–COMPUTER INTERACTION

Advanced technologies are emerging as a new interaction paradigm through HCI in different applications, to provide multi-modal interactions, interactive recommendations, and customization to co-create more value for information systems. In an interactive industrial application, advanced features involve interactions and explores users' preferences to improve UX that demands an extensive analytical tool for professionals in decision-making.

Integration of emerging technologies with information systems has been considered a driving force behind the shaping of a paradigm shift in HCI, which embeds cognitive, social, technological, and environmental effects in achieving organization goals, and professionals need to follow a comprehensive decision-making method to tackle uncertainty. Firstly, these interactions provide new opportunities for organizations through interaction capabilities, moving forward to achieve an interaction paradigm. However, it creates uncertainties with uneven circumstances. Professionals need to address new ways of analytical methods to deal with HCI in the heterogeneous environments. In Industry 4.0, service innovation is transforming towards HCI paradigm, which engages the capabilities of emerging technologies in different industrial applications, but there are challenges to rational decision-making in critical situations to maintain sustainable growth.

This chapter clearly outlines the consequences of HCI, particularly addressing situations and their impacts within key phases of data-driven decision-making. As HCI expands the management capability of an organization, it explores professional engagements for different industries such as health, manufacturing, transportation, education, government service delivery, retail, agriculture, logistics, and tourism. This includes assessing the impacts of value co-creation, knowledge boundaries, and predictive recommendations that can influence the decision-making process for contextual cases. It demands adopting a new framework for managing uncertainties in critical situations, specially dealing with applications of HCI through advanced technologies. The structured framework can explore the reinforcement process to address uneven situations when applied in various real-world contextual applications. Thus, there is a critical need for professionals to formulate a comprehensive strategy to explore innovative alternatives by reducing uncertainties in large-scale operations and enhance organizational performance.

Information systems through HCI engage innovative alternatives for task execution in an organization, affecting professionals' attitudes to perform decision-making. Moreover, it is explained that their motivation to use a specialized analytical platform like power BI and AI-driven agentic models are related to identify alternatives in executing tasks. Also, fulfilment of expectations and influences from HCI can avoid risks and uncertainties in continuing to use information systems. In an organization, professionals' trust in adopting analytical decision-making methods plays a key role to sustain in uncertain situations for business continuity with maximum organizational benefits. However, professionals can extend their analytical knowledge to determine intervening effects of HCI in the decision-making process.

This chapter explains HCI through few applications, where emerging technologies perform as invisible actors through collaborative engagement among digital platforms, users, and technological systems. It also describes how professionals can focus on customers' cognitive engagements, UX, behavioural responses, and decision-making in HCI, considering sources of information that may fulfil customers' expectations. The chapter explains few example scenarios to define HCI through practical applications that engages customers' cognitive actions to influence them towards an information system.

6.4 DECISION-MAKING IN HUMAN–COMPUTER INTERACTION

Cognitive systems theory explains how cognitive functions of humans such as perception, memory, reasoning, and problem-solving mechanisms work through HCI in complex adaptive systems (Hollan et al., 2000). It elucidates the human thought process, human intelligence, and reaction mechanisms that subsequently influence professionals' decision-making. This theory states that cognition is not a detached process but emerges with the HCI between individuals, environment, and information systems. This concept was widely applied in areas like autonomous driving, smart logistics, smart farming, robotic arms, and smart factory to analyze how cognitive frameworks can evolve in response to new sensory information and environmental cues with uncertainty. The cognitive systems theory posits that cognitive activities are not limited to an individual but are shared mechanisms through collaborative interactions among people, tools, and systems. This approach plays a fundamental role in designing intelligent systems that enhance decision-making and problem-solving in complex, interactive, and dynamic settings.

In HCI, users interact with machines or digital platforms and engages sensory information that influences users' decision-making. There are two folds in assessing perceptions like users those who interact with platforms and another group is professionals those who work for organizations in making decisions. For example, in the retail industry, professionals who work in a retail company make their decisions to support customers, vendors, suppliers, solution providers, etc. In contrast, customers who interact with retail shops through offline and online channels, expect quality products at reasonable cost. Therefore, multiple groups of stakeholders have to make decisions using the same platform, which incorporates HCIs in a heterogeneous environment. Decision-making perspectives from stakeholders are crucial for business continuity and growth of industries.

In particular, users engage with information systems through multi-modal interactions, where platform providers try to accommodate sensory information in delivering digital services to provide maximum user comfort. Nowadays, systems engineers develop human-centred design in implementing information systems, where they try to empathize users' perceptions before the systems' commercialization. Therefore, platform developers, UX designers, engineers, IT professionals, solution providers, and device manufacturers can extend their journey mapping towards the analytical process and make rational decisions in every step to achieve sustainable growth in business continuity.

In an integrated and complex adaptive system, professionals' decision-making is more challenging because of its critical situations, particularly to identify alternatives in HCI. Professionals' engagement in the decision-making process is an iteration process to make the service deliberations with optimum outcomes. Conversely, the integrated system involves cognitive and emotional engagements that afterward influence users' behavioural responses through decision-making. Both the decision-making perspectives are critical in achieving organizational goals, particularly to maintain sustainable business growth.

Customers perceive UX after interacting with platforms, and they can make decisions such as purchasing products, continuing to use platforms for services or goods, engaging in word of mouth or recommending someone to use platforms, or opting in to digital services. Interaction capabilities through advanced technologies can impact customers' behavioural responses. In this case, customers are individually affected for their decisions, and they get consequential effects for their decision-making. In contrast, retail owners, professional staff, and platform developers are another group of stakeholders that makes decisions to support retail business. In HCI paradigm, they have to make decisions in implementing critical systems that can impact the whole business, and it is not an individual case, rather it affects collectively.

Furthermore, interactivity with an information system provides sense of information that can confirm the fulfilment of customers' expectations, influencing their decision-making process towards responses. In contrast, professional staff who are responsible for supporting business growth, have decision-making processes that may differ from those of individual customers. UX designers and researchers make adequate decisions in every step from low-fidelity to high-fidelity prototype UX design. Then, application developers need to make decisions to develop digital applications like e-shop mobile app. After the commercialization of digital service in the real world, sales, marketing, and business analyst team make instant decisions considering market scenario, particularly in the heterogeneous environment like multi-vendor applications. They need to deal with different sub-systems like payment system, ERP (Enterprise Resource Planning), logistic apps, ITIL (Information Technology Infrastructure Library) & ITSM (IT service management) like ServiceNow, business analytics platforms like Power BI, etc., and it creates complex situations for professionals in making rational decisions.

6.5 USE CASE SCENARIO

6.5.1 Smart Retailing

Smart retail accommodates human computer interaction, where users interact with mobile devices to engage in shopping virtually and emerging technologies play a significant role to improve UX through virtual product presentation and interactive recommendations without attending the real shopping places (Pantano & Dennis, 2019). Multi-modal interactions with both products and peoples are contemplated in the virtual environment in the smart retailing. The smart retail platform is much easier to access for the end-user for comprehensible, rapid response time, technologically adopted features, and robust enough to gather information in one platform. User modelling in the HCI like in smart retailing, includes unique features like openness, suitability, and explainability to identify the influence of any interactive system on users' perception.

HCI is transforming into a paradigm shift that engages multi-modal interactions from different perspectives and forms an ecosystem to get interactions among users, products, devices, networks, and platforms. In a smart retail platform, emerging technologies engage multi-modal interactions to improve UX and explore innovative features to fulfil users' preferences and choices. This requirement has increased due to rapid changes in user expectations for heterogeneous environments in HCI.

Retailers are adopting advanced technologies such as AR/VR, AI, and IoT to provide interactive and immersive experiences through new interaction features and allows customers to perceive intrinsic value. Recently, few online retail shops incorporated AR view into their mobile platform that develops an added information system to enhance UX through multi-modal interactions. The rapid development of an interactive e-commerce application incorporates AR that transforms a new way of virtual interactions that can alleviate the need for touch and also can allow users to avail virtual try-on features. There are few retailers such as Sephora, Ray-Ban, IKEA, Nike, etc. that accommodate AR view into their existing mobile apps that provide immersive experiences to view digital products through virtual interactions. These are appropriate examples of smart retailers who introduced HCI in the retail industry.

AR in smart retailing co-creates immersive values to improve users' attitude, trust, and satisfaction through the confirmation of their initial expectations from the platforms. AR can accommodate the recognition, detection, and rendering capabilities of virtual products in the smart retail platform using deep neural networks (DNN) through HCI that may engage mental imagery, sense of immersion, sense of presence, and mental imagery towards virtual products and interactive platforms (Kabir & Kang, 2024). The smart retail platform incorporates real-time product tracking, scale measuring, and AR positioning features into existing applications using a DNN of AI, visual-inertial odometry (VIO), and geospatial recognition that creates user trust through forming mental imagery of virtual products. AR engages sensory information to develop users' subjective norms that positively affect users' trust in the smart retailing environment.

Interactions enabled by AR in smart retailing empower users with an intense sense of control by allowing them to place and visualize virtual products within real-world environments. These immersive interaction capabilities encourage adoption through social influence, as peer groups are more likely to recommend and legitimize continued use of AR-based platforms. Such interaction features therefore strengthen subjective norms in smart retailing contexts, where recommendations from peers reinforce users' intentions to engage with digital channels. In parallel, AI-driven recommendations introduce a novel form of persuasion by shaping users' perceived social norms, influencing attitudes and fostering trust in the platform.

In addition, online reviews play a critical role in shaping user perceptions and decision-making in smart retailing. Acting as electronic word of mouth (eWOM), reviews provide interactive feedback that supports virtual evaluation prior to and after product use. Within human–computer interaction, users' behavioural intentions are significantly influenced by online reviews disseminated across multiple digital channels. These reviews serve as a rich external information source, offering social validation and confirming UX, particularly by reinforcing positive perceptions following initial engagement with a product or service.

Smart retailing offers an innovative approach, considering few sensory factors related to product virtualization, sense of immersion, mental imagery, and cognitive attachments through HCI. It also provides simulated physical control (SPC) and product embodiment to mediate the spatial presence of a virtual product in physical space that develops situated experience. The situated experience through multi-modal interactions gives immersive values that influence users to perceive spatial presence. By introducing this immersive feature, retailers can allow online users to check their virtual products and give a proper feeling to retain them with online shopping. This is a good mechanism to make a sustainable growth for the retail industry by increasing user satisfaction with the confirmation of users' expectations from the online platforms.

In smart retailing, AI enhances interaction capabilities by offering new sources of information and encompasses predictive analytics through algorithms, where it can involve more HCIs. AI is an emerging technology that adopts intelligent decision-making processes, and it can predict user requirements by extracting insight. Smart retailing uses affective AI through adaptive and personalized recommendation systems for predictive DA using past purchasing behaviour. The AI-driven recommendation is considered predictive information regarding user characteristics based on previous activities, and this external source of information influences users' subjective norms. AI integrates interaction processes between AI agents and platforms in HCI, delivering intelligent, responsive, and interactive services to end-users. Moreover, advanced algorithmic models enhance human–AI interaction in smart retail platforms and explore potentialities through multi-modal interactions, where those interactions allow product virtualization and predictive recommendations.

In the digital era, smart retailing executes AI algorithms to integrate virtual product recognition, detection, and inspection using DNN, VIO, and geospatial recognition through HCIs, and allows customers to interact with sensory environments. AI engages a new way of interaction capabilities in smart retail platforms by accommodating virtual positioning, experiential information, and virtual try-on features to check the virtual product in a physical space. Further, tangible augmented reality (TAR) through AI allows digital platforms to adapt a spatial registration process by changing the orientation of a desired space, and gesture-based interaction can use front and back mobile cameras to position the products through in-air gestures. Therefore, AI functions as an invisible actor in delivering new features in the smart retailing through real-time product tracking, scale measurement, and product virtualization.

In the smart retailing context, it reveals that users' interaction with digital platforms is essential. In an AR environment, users engage with more sources of information to try on virtual products in physical spaces and check product quality. However, explainable and predictive AI in smart retail platforms can predict users' preferences and deliver contextual interactive recommendations using post-hoc explanation and predictive techniques. Smart retailing creates an interactive environment where retailers can accommodate advanced technologies like AR, VR, AI, IoT, etc., and users can avail immersive and interactive features from digital platforms.

6.5.2 Agentic AI

Agentic AI engages predictive analysis to deliver assistive services to users by extracting insights from all users' touchpoints in HCI that can deliver personalized recommendations to users. In the agentic AI platform, AI applies predictive and explainable AI models to extract valuable insights and experiences from touch points that may deliver a precise recommendation by predicting user preferences and responses (Russell & Norvig, 2021). Furthermore, predictive analytics in agentic AI can make it possible to understand users' behaviour, choices, and past purchasing preferences to deliver interactive responses and recommendations to users. Moreover, digital platforms integrate agentic AI to accommodate personalized messaging and recommendations through predictive DA, analyzing users' meaningful feedback, behavioural responses, and past experiences. Platform developers can integrate their preferences and pre-trained data to address business propositions and service value chain in delivering AI-driven intelligent services to end-users. Professionals need to make rational decisions during service policy design of agentic AI considering business perspectives. Auto-generated personalized recommendations and responses in agentic AI introduce a new way of delivering interactive services to users through HCI by analyzing valuable perceptions, preferences, and service predilections, which can function as an invisible actor towards decision-making.

It is evident that the pre-training and post-training data in agentic AI are based on factual knowledge obtained from HCIs. These interaction capabilities assist an autonomous information system to prevent, detect, and mitigate uncertainties by analyzing contextual information. This capability gives an efficient decision support system and encourages data-driven innovation in the decision-making process. This delivers credible decision-making through data-driven innovation and effective detection and mitigation strategies. Agentic AI based on large language models (LLMs) can analyze pre-training data, implement governance methods, and utilize research and business development (R&BD) mechanisms with contextual information to prevent AI hallucinations and foster data-driven innovation in the decision-making process to avoid uncertainties.

For instance, hallucinations from LLM-based agentic AI may influence digital platform outputs, such as reinforcing gender stereotypes and ethnic biases. When these hallucinations involve misinformation, like fake news or rumours, it can mislead users in users' decision-making based on faulty recommendations. Agentic AI can employ advanced features to explore the impacts of AI hallucinations through three sequential steps: (i) trust erosion, (ii) inaccurate outcomes, (iii) dissemination of misinformation, and (iv) dissemination of erroneous content. This process is essential for identifying hallucinations and requires detecting the root causes to mitigate inconsistencies and avoid unwanted outcomes. Using agentic AI, organizations can develop mechanisms for validation, intervention, and activation to mitigate the spread of misinformation and unwanted content through their platforms. This chapter highlights the increasing need for effective LLM-based agentic AI to avoid uncertainties like AI hallucination through prevention, detection, and mitigation, particularly in the realm of data-driven innovation (Kabir, 2026).

This requirement has been demanded to deploy agentic AI due to the rapid adoption of HCI in different industries. In the age of AI, use of agentic AI includes information verification mechanism towards data-driven innovation. AI can function like humans in processing information through agentic AI mechanisms that may explain self-efficacy towards decision-making by describing the capability to perform, reinforced learning, adaptation, and HCI capabilities.

The agentic AI evolves AI's capabilities through reinforced learning, improvements of algorithms, and re-adjustments based on HCI, particularly for autonomous systems. This dynamic nature of agentic AI means that its developers and professionals must be able to manage these constant changes in operational environments to make their decisions. Effective management of agentic AI requires professionals to have the dynamic managerial capability knowledge to oversee its learning, decision-making policy, and integration techniques into existing information systems. This is crucial for aligning agentic AI with evolving business goals or societal demands.

Agentic AI can explore the reinforcement process to avoid unwanted circumstances, inconsistencies, and uncertainties with pre-training data when executed in the real-world scenarios. Therefore, there is a demand for a comprehensive decision-making approach to adopt large-scale human-computer interactive platforms. The decision-making framework must remain consistent with the proper validation process through pre-training data, post-training data, and contextual HCI-based information. Furthermore, the framework can incorporate validation, intervention, activation, inspection, estimation, evaluation, judgement, data, methods, training, governance, and R&BD as essential indicators for defining the prevention, detection, and mitigation of uncertainties in the context of LLMs-based agentic AI, aiming to achieve data-driven innovation within an organizational structure.

6.5.3 Next Generation Wireless Network

Network providers, service providers, and application providers are facing new challenges to cater to the needs of a radical shift from network quality to human-centred design through UX assessment techniques that not only consider key parameter indicators (KPIs), but also evaluate QoE, UX, QoD, and quality of machine intelligence. This requirement has emerged due to abrupt changes in user expectation by expanding heterogeneous network environment through human-to-machine or machine-to-machine communications. In the 5G era, there is a scope to accommodate intelligent DA to evaluate user interactions and correlate them with the network data. In next generation wireless network, the whole mobile network ecosystem is reframed through enhanced mobile broadband (eMBB) services and UX would become a critical area to maintain uninterrupted services, considering both the services and contextual scenarios.

In the last decade, network providers have concentrated to increase QoS through maintaining KPIs in the network side. Currently, network providers have incorporated QoE-based quality assurance system to get feedback from end-users through maintaining KQIs. The communication industry is going faster to accommodate smart technologies to assess UX considering both services and scenarios (Odarchenko et al., 2020).

Over the last few years, there has been a significant increase in the number of services in mobile networks. This trend has forced network providers to change their network management processes to ensure adequate QoE, instead of adequate QoS. As a result, customer experience management (CEM) is now a critical task for all the relevant providers, who demand tools for QoE monitoring on an individual user basis. The ever-increasing demands of users related to high storage, fast computation, and processing lead towards the innovation of 5G services.

QoS typically focuses on telecommunications services, whereas QoE covers a much broader domain, which sometimes does not even involve telecommunications, e.g., HD video played in a home theatre. QoS deals with performance aspects of physical systems. QoE deals with the users' assessment of system performance, as coloured by context, culture, the users' expectations with respect to the system or service and their fulfilment, socio-economic issues, and psychological profiles, among other factors. QoS has a very technology-oriented approach, and it relies on analytic approaches and empirical or simulative measurements. QoE, on the other hand, requires a multi-disciplinary and multi-methodological approach for its understanding.

QoE research is rooted in research activities that aim to determine the impact of service characteristics on the perceptions of human users. The QoE introduces a concept that defines a degree of end-user's "delight or annoyance" when using a product or service, particularly mobile services in the communication industry. In the past, QoS was the key performance metric of every network because it could be reliably measured and helped network engineers to improve network services. However, QoS parameters reflect only the network performance, which does not necessarily indicate the trust and satisfaction of end-users. Hence, the estimation model of user satisfaction in terms of QoE by using neural network approach was proposed. The QoE model assesses user satisfaction resulting from the QoS parameters instead of directly measuring the mean opinion score (MOS) from users. The novel concept of Fifth Generation (5G) cellular networks brings a combination of a diversified set of devices and machines with significant improvement in a unique way compared to previous technologies. To broaden the mobile user's experience, 5G technology provides the opportunity to meet the people's potential necessities for efficient communication. According to the wireless network technologies analysis, the next-generation cellular network is expected to deliver each user with ultra-reliable connectivity via multiple devices anywhere and anytime, which demands advanced UX analytical tools to maintain uninterrupted services.

The next-generation network is considered a multi-tier heterogeneous network (HetNet) because it supports various wireless devices in each tier. Each tier consists of small gadgets to heavy machine-type wireless equipment and sends data requests directly to its respective small base station (SBS), which requires low power for transmission and minimized energy consumption. Supporting of the required level of user satisfaction in high-quality services has complex character and presumes the system of organizationally technical and socio-economic events for reduction of the achieved service quality level to existing, incipient, or prescribed needs of subscribers.

The motivation is extended by the futuristic demand for UX in the next-generation wireless networks, i.e., connectivity anywhere (wherever needed), anytime (whenever needed), for anyone (whoever needs it), or for anything (whatever needs it) to cater to eMBB use cases. Customer engagement hub, customer journey analytics, and graph analytics for customer experience (CX) are depicted as peak of inflated expectations in information systems engineering. Further, multi-experience analytics and customer psychographics are depicted as innovation trigger for the next projections. Moreover, small-cell and cloud-native CSP infrastructure are depicted as peak of inflated expectations and private 5G, 5G-advanced, 5G network security are depicted as innovation trigger in the future of communication service providers (CSP) networks infrastructure.

Advanced research on network engineering explains about the UX for different application scenarios and its associated KPIs of 5G services. There are diverse application scenarios and KPIs, that varied for immersive 5G services accordingly, not just the peak data rate and average spectral efficiency as required in the earlier generations. For stream-type services, data rate of 8K resolution (3D) would be 96 Gbps (without compression) and would be 960 Mbps (with compression by 100x). The delay is tolerated up to 5~10 ms for an interactive service, so fast that users would not be aware of it. Enhanced reality and smart technologies require random movement of hyper-realistic media, meaning that the systems have to provide high-speed communication both on the downlink and the uplink. Application scenarios and immersive services are highly coupled in 5G wireless system. Therefore, a comprehensive and advanced decision support system is required to assess UX for next-generation wireless network.

6.5.4 Ubiquitous Society

Digital services are transforming into a ubiquitous society, where a global megatrend establishes a converged platform that enhances communication services through the explosion of data traffic, connected digital devices, cloud-based infrastructure, hyper-realistic media, and heterogenic data sources. In this chapter, we explain how this megatrend reshapes the decision-making process through HCI. The ubiquitous society accommodates immersive, intelligent, omnipresent, autonomous, and predictive services that interact between human and machines or machines and machines in delivering optimal benefits. Next generation services in the ubiquitous society engage integration being provided using AR/VR, telepresence, 3D holograms, AI, cloud computing, IoT, and brain-computer interface that transmit massive content streaming with higher bandwidth, low latency, and speedy data rate.

AR/VR requires a replica site of user with sufficient image/video quality, viewing angle, input interface, and head-mounted display (HMD). Telepresence services can give users a more realistic experience using VR and 3D holograms. 5G networks provide high bandwidth, high transmission rates, and low latency. Moreover, the development of digital technologies increases the demand of massive content streaming through 8K ultra-high definition (UHD) content, multi-view interactive 3D data, personal multimedia broadcasting, and massive content sharing. Multidimensional and hyper-realistic media data offers information that can be seen, heard, and felt through the five senses. Hyper-realistic holographic technology will provide realistic services, and it requires a bandwidth of several terabytes per second.

Moreover, the extended reality (XR) environment caters to immersive 5G services to interact with both the physical and virtual worlds in an imaginary space. XR is defined as an umbrella term to include VR, AR, and MR. VR delivers computer-generated images, videos, and sounds using a HMD, in which the user can control objects and move around through haptic controllers. AR accommodates the real world with digital auto-generated information and enhances the real environment with different fields of view (FOV). XR technologies provide immersive experience in different scenarios of use cases.

The ubiquitous society needs to maintain uninterrupted services due to its nature of service deliberations. For example, autonomous driving vehicle communicates data through intelligent traffic systems and continuously share information to other devices like mobile handset, rechargeable station, and sensor devices. In a ubiquitous environment, perceptive network through AI allows next-generation networks to avail human-like qualities to self-manage and adapt the network in the changing environment of human and machine expectations. The perceptive network functionalities are defined as a tactile network that considers human behavioural characteristics such as: (i) foreseeing human and machine requirements and actions; (ii) creating a self-adaptive mechanism to identify uncertainty in the inherent heterogeneity; and (iii) offering a better QoE and improving system performance. To be considered human-centric, the ubiquitous system thus provides proactive accommodation of human and machine behaviour, which refers to the anticipation of human and machine intent and characteristics, allowing the digital services and the communication systems to adapt to it proactively. Human and machine-centric methods and self-management mechanisms can facilitate negotiation between the stakeholders currently engaged in envisioning the ubiquitous society, including machines. Therefore, a challenge is to assess the subjective measures of human and machine experience in the decision-making process. This chapter highlights the implications of human-centric design interpretations on the decision-making methods for the ubiquitous society.

In a ubiquitous environment, QoE optimization in next generation communication networks introduces Radio-Access Network configuration (RA-NC) optimization modelling on the impacts of QoE-oriented optimization. An end-to-end QoE optimization model is required to maintain uninterrupted digital services in the ubiquitous society.

Further, professionals can deploy an experience delivery network (XDN) for ubiquitous services, which is cloud-based solution and can be applied in different contextual scenarios. The XDN includes 5G-driven multi-access edge computing (MEC) which demands ultra-low latency and ultra-high bandwidth and reducing the computational load of end-users. Immersive media (8K 3D video) requires 300 Mbps bandwidth in uplink and downlink with 10 ms of end-to-end latency. Immersive media 360 (IM 360) network layout is approached for an end-to-end edge-based solution with ultra-high resolution of 8K 360 degrees video, live streaming, and broadcasting. IM 360 is composed of three main modular functions. Media control function (MCF) controls authentication, session management, and cluster management. Media ingestion function (MIF) controls media ingestions (upstreaming) from different capture devices. It supports various protocols like RTMP, RTSP, SRT, various video codecs such as H.264, HEVC, VP9, etc. Media distribution function (MDF) manages media distribution with common 2D video and view-port independent streams. For viewport-dependent streaming, the MDF revisits the bitstream (HEVC) and re-packetizes RTP streams. A room has one MIF as content provider, and multiple MDFs as content consumers. 5G and edge computing are engaging new experiences through an experience delivery network (XDN) for immersive media.

For ubiquitous services, a DA-driven customer experience management (CEM) framework can be executed for decision-support system to maintain QoE for digital services. In a machine communication environment, key parameters are often time-sensitive and sensor-based data. QoE can be assessed through subjective evaluation such as experience, emotions, expectations, machine intelligence, and objective evaluation including both technical and non-technical factors. Multi-operators and multi-vendors are operated in the heterogeneous environment to share network capacities as a vertical industry. The CEM customizes service delivery propositions in the perceived user-centric network with proactive data analysis that can maintain the quality-of-service deliberations. Human and machine-centric service requirements can be analyzed through time and location-based awareness to assure modular and adaptive framework to support rational decision-making process.

The DA-driven CEM framework includes key functional layers such as execution, recommendation, insight, network, and data collection. The framework is designed to integrate diverse data from various sources like user-terminal, cloud services, radio access network, core switch data, etc in the data collection layer. Then, value is extracted for CEM (QoS-to QoE) from the metrics using DA in the insight layer. Network statistical projection, identity trends, and new recommendation rules are provided in the recommendation layer. The execution layer deals with the interfaces towards networks, handles user functions, and establishes caching servers at remote locations. Professionals can follow this framework to maintain UX for the ubiquitous services.

In the ubiquitous era, decision-making would be becoming an emerging issue to mitigate some challenges like:

- Heterogeneous networks: Multi-operator and multi-vendor environment co-exist in the ubiquitous environment. Both human and machine sensitive network need near zero latency and uninterrupted service reliability to enhance UX.
- Application and content: The UX model has to cater for versatile applications through numerous types of content including machine-type communication (beyond basic features of communication services).
- End-user engagement: Both machines and humans function as prosumers in the service delivery mechanism and the ubiquitous society is going towards human and machine-centric network to fulfil end-terminal requirements.

The QoE-oriented optimization for decision-making framework can explain QoE variable, QoE objective, QoE constraint, and optimization strategy. In the model, QoE objective depends on the variables and need to be optimized through strategy. QoE variables denote some subjective and objective factors including KPIs for the network quality (throughput, jitter, etc.), context property to evaluate users' profile (emotion, psychological needs, user preference, etc.) and terminal dependent factors (terminal capabilities, screen size, and energy efficiency). QoE objective is a model to evaluate QoE and also enhances the optimization of QOE. QoE objective is derived as a measurement process by surveying end users' experience. QOE feedback is determined to achieve QoE objective through estimating QoE of both subjective QoS and objective context property.

6.6 CONCLUSION

Contemporary decision-making within human–computer interaction (HCI) increasingly depends on the integration of personalized insights, interconnected systems, and agile methodologies capable of responding to critical and dynamic situations. In heterogeneous digital environments, shifting user expectations and evolving interaction capabilities continually redefine key parameters, demanding comprehensive models that can evaluate decision contexts and viable alternatives. HCI strengthens the connective fabric between human and non-human systems, enhancing the capacity of information systems to support informed decisions. As HCI evolves, it is becoming more sustainable and transformative through the incorporation of intelligent systems and modern technologies, which together elevate the perceived quality of UX and create an enabling environment for effective decision-making.

In conclusion, the emergence of advanced technologies has introduced new paradigms of interaction, characterized by multi-modal, interactive, and context-aware experiences across diverse applications. This chapter has demonstrated how cognitive functions shape decision-making within HCI, illustrated through use cases such as smart retailing, agentic AI, next-generation wireless networks, and the vision of a ubiquitous society. These examples reveal how emerging technologies increasingly operate as invisible yet influential actors, facilitating collaborative engagement among users, digital platforms, and intelligent systems. The chapter underscores the importance for professionals to prioritize cognitive engagement, UX, behavioural responses, and decision outcomes, ensuring that information sources and interaction designs align with and fulfil evolving user expectations.

BIBLIOGRAPHY

Anchuen, P., Uthansakul, P., & Uthansakul, M. (2016). QOE model in cellular networks based on QOS measurements using Neural Network approach. In *2016 13th International Conference on Electrical Engineering/Electronics, Computer, Telecommunications and Information Technology (ECTI-CON)*, IEEE, 1–5. https://doi.org/10.1109/ECTICon.2016.7561318.

Barakabitze, A. A., Barman, N., Ahmad, A., Zadtootaghaj, S., Sun, L., Martini, M. G., & Atzori, L. (2020). QoE management of multimedia streaming services in future networks: A tutorial and survey. *IEEE Communications Surveys and Tutorials*, 22(1), Article 8930519. https://doi.org/10.1109/COMST.2019.2958784.

Chen, C., Zhang, K. Z. K., Chu, Z., & Lee, M. (2024). Augmented reality in the metaverse market: The role of multimodal sensory interaction. *Internet Research*, 34(1), 9–38, doi: 10.1108/INTR-08-2022-0670.

Costa, R. L., Viana, A. C., Ziviani, A., & Sampaio, L. N. (2021). Tactful networking: Humans in the communication loop. *IEEE Transactions on Emerging Topics in Computational Intelligence*, 5(1), Article 9285188. https://doi.org/10.1109/TETCI.2020.3039520.

Du, Z., Liu, D., & Yin, L. (2017). User in the loop: QoE-oriented optimization in communication and networks. In *2017 6th International Conference on Computer Science and Network Technology (ICCSNT)*, IEEE, 420–424, doi: 10.1109/ICCSNT.2017.8343731.

Gacanin, H., & Wagner, M. (March/April 2019). Artificial intelligence paradigm for customer experience management in next-generation networks: Challenges and perspectives. *IEEE Network*, 33(2), 188–194, doi: 10.1109/MNET.2019.1800015.

Hammer, F., Egger-Lampl, S., & Möller, S. (2018). Quality-of-user-experience: A position paper. *Quality and User Experience*, 3(1), Article 9. https://doi.org/10.1007/s41233-018-0022-0.

Haroon, M. S., Abbas, Z. H., Muhammad, F., & Abbas, G. (2020). Analysis of coverage-oriented small base station deployment in heterogeneous cellular networks. *Physical Communication*, 38, Article 100908. https://doi.org/10.1016/j.phycom.2019.100908.

Hassenzahl, M., Diefenbach, S., & Göritz, A. (2010). Needs, affect, and interactive products – Facets of user experience. *Interacting with Computers*, 22(5), 353–362. https://doi.org/10.1016/j.intcom.2010.04.002.

Hollan, J., Hutchins, E., & Kirsh, D. (2000). Distributed cognition: Toward a new foundation for human-computer interaction research. *ACM Transactions on Computer-Human Interaction*, 7(2), 174–196. https://doi.org/10.1145/353485.353487

Huta, V., & Waterman, A. S. (2014). Eudaimonia and its distinction from Hedonia: Developing a classification and terminology for understanding conceptual and operational definitions. *Journal of Happiness Studies*, 15(6), 1425–1456. https://doi.org/10.1007/s10902-013-9485-0.

I, C.-L., Han, S., Xu, Z., Wang, S., Sun, Q., & Chen, Y. (2016). New paradigm of 5G wireless internet. *IEEE Journal on Selected Areas in Communications*, 34(3), Article 7399671. https://doi.org/10.1109/JSAC.2016.2525739.

Jumisko-Pyykkö, S., & Vainio, T. (2010). Framing the context of use for mobile HCI. *International Journal of Mobile Human Computer Interaction*, 2(4), 1–28. https://doi.org/10.4018/jmhci.2010100101.

Kabir, Z. S. (2026). Explaining Hallucination of Artificial Intelligence in Data-Driven Decision-Making. In Z. Kabir (Ed.), *Innovative Decision-Making in Engineering: The Role of Cognition, Heuristics, and Human Factors* (pp. 85–112). IGI Global Scientific Publishing. https://doi.org/10.4018/979-8-3373-5702-7.ch004

Kabir, Z. S., & Kang, K. (2024). The impact of augmented reality through user-platform interactions towards continuance intention with the effect of user generation. *Information (Basel)*, 15(12), Article 758. https://doi.org/10.3390/info15120758

Odarchenko, R., Abakumova, A., Bondar, S., Bogachuk, Y., & Maratkyzy, K. (2020). Advanced method for QoE evaluation and improvement in modern cellular n`etworks. In *2020 IEEE International Conference on Problems of Infocommunications.* Science and Technology (PIC S&T), 289–294. https://doi.org/10.1109/PICST51311.2020.9468036.

Pantano, E., & Dennis, C. (2019). *Smart retailing: Technologies and strategies*, Springer International Publishing. https://doi.org/10.1007/978-3-030-12608-7.

Philip Chen, C. L., & Zhang, C.-Y. (2014). Data-intensive applications, challenges, techniques and technologies: A survey on Big Data. *Information Sciences*, 275, 314–347. https://doi.org/10.1016/j.ins.2014.01.015.

Reed, L., Li, C., Ramirez, A., Wu, L., & Walker, M. (2022). Jurassic is (Almost) all you need: Few-shot meaning-to-text generation for open-domain dialogue. In S. Ultes, H. Li, & S. Stoyanchev (Eds.), *Conversational AI for natural human-centric interaction* (Vol. 943, pp. 99–119). Springer. https://doi.org/10.1007/978-981-19-5538-9_7.

Russell, S., & Norvig, P. (2021). *Artificial intelligence: A modern approach* (4th ed.). Pearson.

Shen, G., Dai, J., Moustafa, H., & Zhai, L. (2021). 5G and Edge computing enabling experience delivery network (XDN) for immersive media. In *2021 IEEE 22nd International Conference on High Performance Switching and Routing (HPSR)*, 1–7. IEEE. https://doi.org/10.1109/HPSR52026.2021.9481809.

Siddiqui, M. U. A., Qamar, F., Ahmed, F., Nguyen, Q. N., & Hassan, R. (2021). Interference management in 5G and beyond network: Requirements, challenges and future directions. *IEEE Access*, 9, Article 9405679. https://doi.org/10.1109/ACCESS.2021.3073543.

Wobbrock, J. O., & Kientz, J. A. (2016). Research contributions in human–computer interaction. *Interactions,* 23(3), 38–44. https://doi.org/10.1145/2907069

Zikria, Y., Kim, S.W., Afzal, M., Wang, H., & Rehmani, M. H. (2018). 5G mobile services and scenarios: Challenges and solutions. *Sustainability*, 10, 3626, doi: 10.3390/su10103626.

7 Decision-Making Tools for Professionals in Engineering Practices

7.1 INTRODUCTION

Making decisions involves choosing from available options to achieve a goal and weighing their pros and cons is incredibly challenging for professionals. Decision-making is like an optimization process, selecting the best solution from available options. Identifying and differentiating between alternatives and selecting the best option can be assessed by using decision-making tools. Professionals can measure and evaluate effectiveness to perform their activities through judgemental decision-making.

Deterministic decisions have certain outcomes, while stochastic decisions involve evaluating probabilities and are more common in real-life situations. Decisions are made with the purpose of achieving specific goals. There can be various analytical methods to achieve these goals, which can either complement or contradict each other. It is crucial to keep in mind that each decision is driven by its own unique set of goals, arranged in a specific hierarchy.

To make effective decision-making in an organization, identify the goal hierarchy, assess available analytical tools, involve the right groups, recognize knowledgeable individuals, assign a manager to define the problem, form a team, allocate resources, and use analysis results to reach a rational decision. Various analytical techniques can support professionals to comprehend a situation. These decision-making tools include brainstorming, mind mapping, influence diagram, SWOT, decision tree, and stochastic analysis. However, it is essential to note that these methods do not directly solve the problem at hand. Critical thinking and analysis are necessary to find a practical and effective solution to make rational and judgemental decisions.

7.2 DECISION-MAKING TOOLS

Providing better insights through decision-making tools, highlighting alternatives, giving feedback, and aiding analysis to make strategies that improve decision-making. Expert decision-support systems analyse human expertise and contextual information in limited domains, identifying critical factors and their weightage to analyse tasks and provide optimum benefits considering minimal cost involvement, increased output, and improved quality of service deliberations. Decision-making

DOI: 10.1201/9781003685449-7

tools can enhance professionals' capabilities to make critical decisions that increase the productivity and skills of human expertise by incorporating machine intelligence and generative tools that assist professionals in making rational decisions. The most successful expert systems guide human experts.

Decision-making tools provide an intuitive and efficient way to represent decision-making scenarios. They help decision-makers express their beliefs about uncertainties, preferences for options, and relevant information. Decisions are crucial for professionals in engineering practices, impacting project implementation and product development. Key decision-making tools include information availability and reliability through normative and stochastic modelling, along with methodological approaches like qualitative and quantitative methods, which can accommodate experience, emotional state, contextual information and knowledge to develop rational decision-making. The rational decision-making model, as shown in Figure 7.1, visualizes the process flow of how decision-making tools can support professionals in making rational decisions.

During the decision-making process, information evolves through normative and stochastic decision-making tools and analytical methods like qualitative and quantitative tools, which extract valuable insights and allow professionals to make their rational choices. The key challenge is: How much reliable information is being processed to identify alternatives and choose the best option at each stage for accurate decision-making? Decision-making involves processing critical information, followed by taking rational actions.

Uncertainties can significantly impact decision-making processes. They can lead to analysis paralysis, causing individuals or organizations to overanalyse information and fear making the wrong choice, resulting in missed opportunities. Additionally, uncertainty can induce stress and anxiety, leading to risk-averse behaviour that stifles innovation and growth. Dealing with uncertainty can be challenging, but decision-making tools like decision trees, SWOT analysis, a growth mindset, and gathering diverse perspectives can improve decision-making in a rapidly changing world.

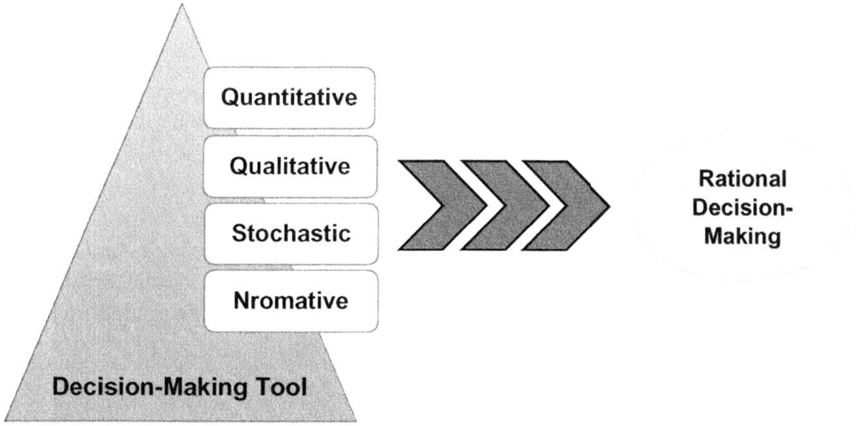

FIGURE 7.1 Rational Decision-Making Process.

Making decisions in engineering practices involves picking the best option from available alternatives to achieve organization goals and weighing each alternative to quantify the decision-making process. The process is like an optimization challenge, choosing the best probable solution considering available resources. Identifying and differentiating between alternatives can be difficult. Evaluating appropriateness can also be a challenging situation for professionals, but professionals can apply decision-making tools to ensure a rational process in making decisions beyond controversies and hesitation.

Decision-making models rationalize how preferences are considered under different situations of information with uncertainties and contextual differences. Broadly, decision-making models can be classified into:

I. Normative or deterministic model – how decisions can be executed to realize optimum level of outcomes.
II. Stochastic model – how decisions can be taken under uncertainty, unpredictability, or unknown context or situations.

This difference is critical for professionals in systems engineering, psychology, ergonomics, management, and industrial operations. In the context of data collection and analysis, decision-making tools can be divided into:

I. Qualitative model – A qualitative model is characterized by its objectives, which connect to explore and interpret valuable insights by analysing data, which generate ideas, rather than numerical values.
II. Quantitative model – A quantitative model aims to determine something related to numerical values and it employs analytical methods to quantify the outcomes.

So, we need some tools, qualitative or quantitative, to identify the best feasible option to execute the task efficiently.

7.2.1 Normative or Deterministic Model

A deterministic or normative model allows professionals to make decisions with certain outcomes. In a traditional process, actors or agents observe the situation by interacting with an environment and take actions according to their perceived experiences. Classical decision-making theories explain the identification of actions, events, consequences, and probabilities in terms of the combination of actions and events.

Normative models define optimal decision rules grounded in logic, mathematics, and economic rationality, assuming fully rational professionals with complete information, predictable outcomes, and consistent input–output relationships, which is why such models are commonly described as deterministic.

7.2.1.1 Brainstorming

Brainstorming is considered to be an initial step to ideate new concepts that can support professionals in making rational decisions. First of all, professionals need to investigate the contextual situation and brainstorm on how many relevant factors are critical to making decisions. They can ask questions to keep the ball rolling and develop principles considering interests. Then, review past experiences to get ideas about how it can relate to current situations. Professionals can zoom out initially to diverge the ideas considering versatilities and then zoom into the identified situational context in making decisions. The choice of relevant information must be justified to identify situations that can give confidence to professionals in executing tasks.

Professionals can brainstorm about vested interests, cognitive involvement, ethical values, and biases. They can fill a gap in what is already known about the topic, by adding some useful information or by applying contextual experiences and methods in new contexts.

A brainstorming mechanism can be applied with few considerations such as broader recognition, information cues, improved insights, divergent and convergent thinking, a multidimensional review process, and creative output extraction.

7.2.1.2 SWOT

A SWOT analysis is an effective tool for allowing professionals to assess risk and for accelerating the process to identify and mitigate high-level risks. It assesses the project or system's implementation from four perspectives: strengths, weaknesses, opportunities, and threats (Phadermrod et al., 2019).

Strengths are considered as a quadrant where an organization excels and frames positively from market competitors. Weakness connects to the quadrant where it relates to any aspect of a business needs, including staff, working capabilities, procedures, culture, or operational output. Having an objective and discerning outlook is crucial to pinpointing these areas.

Opportunities are in the quadrant, where external chances for growth can gain project values, which can be large or small and lead to growth. External threats, such as shortages, supply chain disruptions, volatile stock markets, or product recalls, pose risks to the project.

For example: This chapter delves into the catastrophic Airgas Cantonment Nitrous Oxide Explosion on August 28, 2016. The explosion, which caused significant plant damage and the tragic death of an employee, was due to the improper switch from positive displacement pumps to centrifugal pumps without adequate risk evaluation (CSB, 2017). The centrifugal pumps generated excessive heat, leading to the decomposition of nitrous oxide. The absence of a robust Process Safety Management (PSM) system further exacerbated the situation by failing to identify, assess, and mitigate the associated risks. The Airgas Cantonment Nitrous Oxide Explosion occurred due to the multiple critical decisions taken by both Airgas management and stakeholders. Selection of Nitrous Oxide Transfer Pumps would be considered as a critical situation that could be examined using SWOT analysis (as shown in Figure 7.2).

The decision undertaken brought about major difficulties. The explosion occurred because of the decomposition process of nitrous oxide, which started when centrifugal pumps exceeded their operating temperature. Risk evaluations conducted through combined engineering and safety management teams could have yielded protective

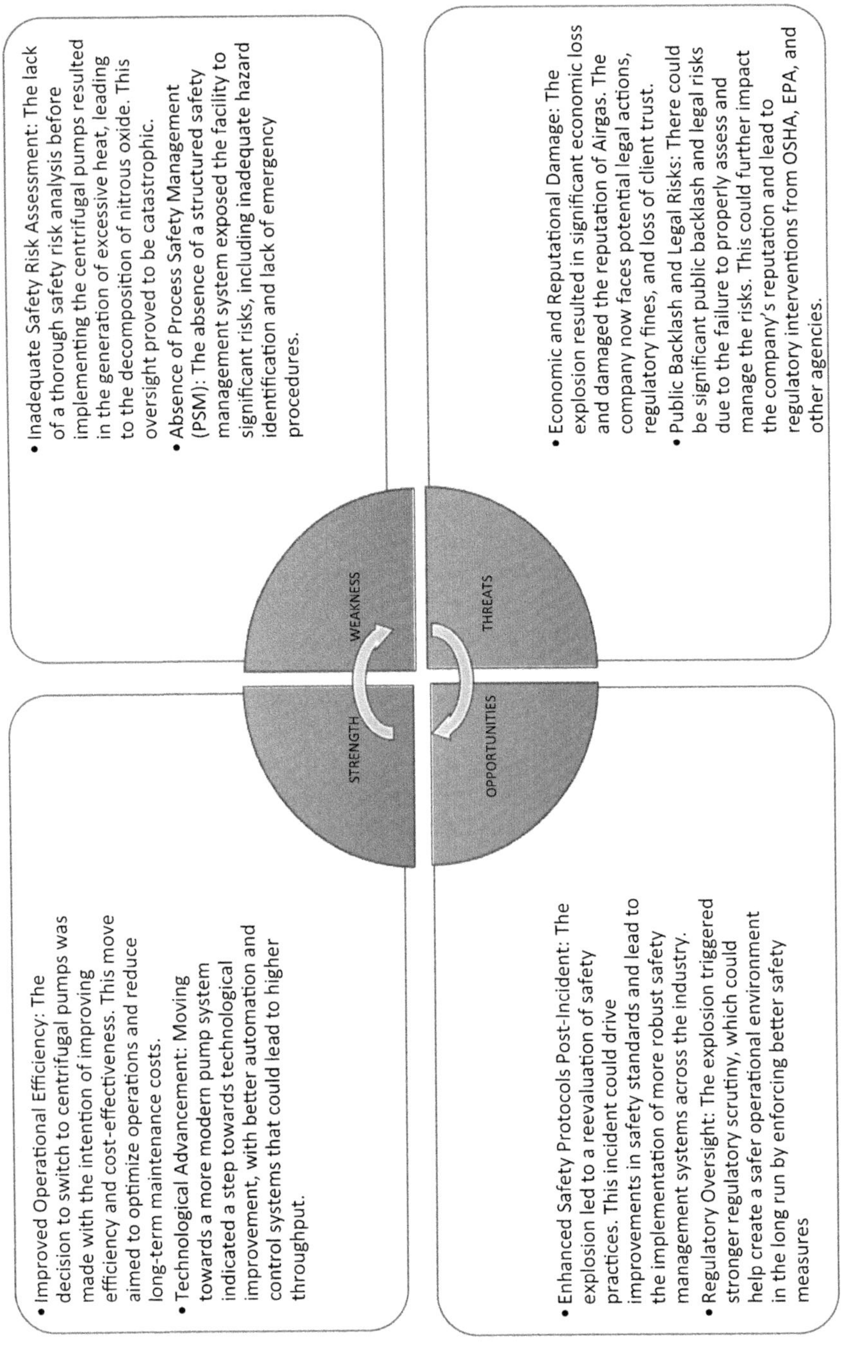

FIGURE 7.2 SWOT Analysis for Airgas Cantonment Nitrous Oxide Explosion.

safety measures for the operation. This improper risk assessment led to Airgas's damaged reputation, along with endangering the safety of its staff. The stakeholders endured negative impacts from inadequate safety oversight because of inadequate decisions that both employees and local communities had to deal with.

7.2.1.3 Decision Tree

Decision tree analysis visually outlines potential outcomes, costs, and consequences of a complex decision based on quantitative data. It uses a graphical representation resembling an upside-down tree with nodes representing decisions or chance events and branches representing actions.

The decision tree process begins with collecting relevant, accurate, complete, and representative data, followed by preprocessing steps such as managing missing values and outliers and applying feature transformations where necessary. Model development then starts at the root node, which represents the entire dataset, and proceeds by selecting the most informative attributes to split the data into child nodes, repeating this process recursively until predefined stopping criteria are met. Decisions are made by traversing the tree from the root to a leaf node according to observed attribute values, with the leaf node indicating the predicted class or outcome, while cross-validation is employed to ensure robust performance on unseen data. Finally, the tree is interpreted by analysing its structure to extract decision rules and identify key features and their influence on the resulting decisions.

Decision matrices and decision trees are analytical tools used to compare alternatives based on expected value or utility, with decision trees providing a structured way to represent complex relationships among objectives, attributes, goals, and values, while different forms of trees are employed to model probabilistic relationships between events and decision choices.

For example: Lack of Safety Management System implementation would be a critical situation for the Airgas Cantonment Nitrous Oxide Explosion. The lack of PSM implementation at the Airgas facility served as a main contributor to the explosion. The facility operated with no established procedures to either evaluate, detect, or regulate hazards within its high-risk nitrous oxide operations (CSB, 2017). The core actors responsible for this situation included Airgas's senior leaders together with its team focused on safety services, who neglected to build an effective security framework. The delay in safety measure deployment stemmed from management's reluctance, which developed from their false belief in controlling the situation. Hazardous working conditions endanger the employees because they receive neither sufficient training nor adequate safety measures from the company. The decision tree can be illustrated (as shown in Figure 7.3).

7.2.1.4 Influence Diagram

An influence diagram presents a functional and informative visual interpretation of the causal and temporal relationships between events and actions, supporting professionals to use as a decision-making tool in the initial stages to make valuable decisions. Influence diagram is a graphical illustration showing different relationships and their impacts that allows professionals to enhance their decision-making. They

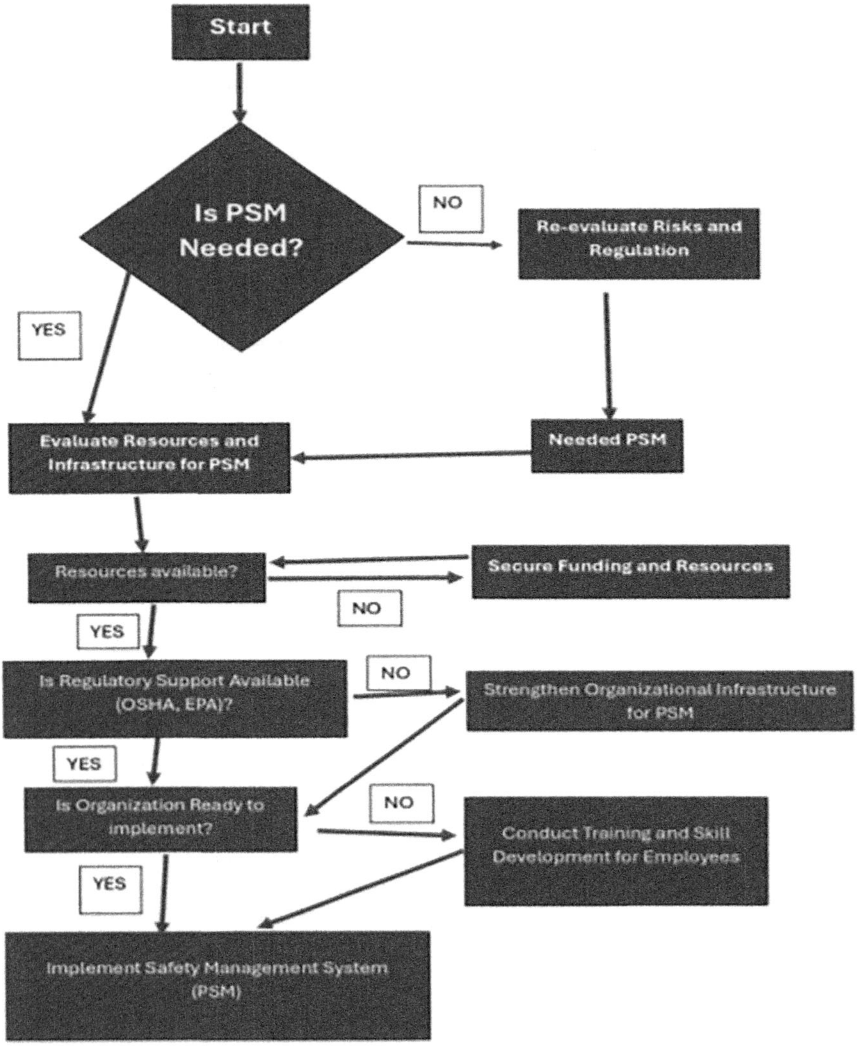

FIGURE 7.3 Decision Tree for PSM.

can depict a visual outline of the relationships between variables and how they proceed for the next consequences.

Conditional relationship is an especially important consideration when drawing an influence diagram that refers to the interdependence of one event with another. Understanding conditional relationships between events can be challenging, especially for those who have limited knowledge of probability. An influence diagram follows conditional relationships and provides a visual representation that supports professionals to identify and analyse relevant factors and their relationships involved in the decision-making process. It clearly indicates how different alternatives, situations, and outcomes are correlated. This yields it simpler and understandable to visualize main factors involved in the decision-making process and how each factor can influence others. An influence diagram is an effective tool in decision-making as it can help recognise the most critical factors that can influence the decision-making.

Influence diagrams use various elements that are represented by distinct symbols, allowing the decision-making problem to be presented graphically. A chance node, represented by a circle (○), denotes an uncertain variable beyond the decision-maker's control whose outcomes are governed by a probability distribution, may depend on other chance or decision nodes, and captures randomness in factors such as market demand, weather conditions, system failures, or customer responses. A value node, represented by a rounded square (▢), captures the objective or payoff of a decision problem such as expected profit, total cost, or customer utility by aggregating the consequences of preceding chance and decision nodes, without assigning probabilities, and typically appears at the final stage of an influence diagram.

A decision node, represented by a square (□), denotes a choice or action under the control of the decision-maker, evaluated by comparing the consequences of alternative options and typically informed by available information, such as whether to launch a product, invest in technology, or select a supplier. A deterministic node, typically shown with a spread sign or double-outlined (function) symbol, represents a variable whose value is precisely calculated from its parent nodes through a known functional relationship, involves no uncertainty or probability distribution, and is commonly used to model equations or logical rules such as total cost equalling fixed cost plus variable cost or revenue being the product of price and quantity.

A relational link, represented by an arrow (→), denotes the direction of influence, dependency, or information flow between nodes in an influence diagram, where arrows into chance nodes indicate probabilistic dependence, into decision nodes show available information, into value nodes represent contributions to utility or payoff, and into deterministic nodes provide functional inputs, with direction indicating influence rather than causality alone. In an influence diagram, the arrow symbol links nodes to express distinct types of weighting as shown in Figure 7.4, where arrows between decision and chance nodes indicate possibility weights, arrows between chance nodes

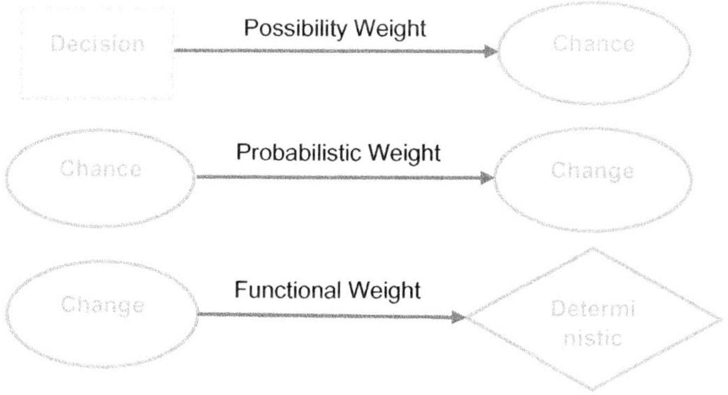

FIGURE 7.4 Weighting Representation in Influence Diagram.

TABLE 7.1
Factor and Node Description for Fire Safety System

Node	Factor	Role in Fire Safety System
Chance	Fire ignition, detection capability, firefighting training, HMI, emergency warning	Initiation of incident
Decision	Safety measures Decisions	Represent decision management actions
Deterministic	damage level	Represent calculated outcomes
Value	Safety, cost, compliance, Shutdown Switchboard	Co-created values to achieve goals

represent probabilistic weights, and arrows from chance nodes to deterministic nodes denote functional weights.

For example: Confirming fire safety issues in an industry, the influence diagram as shown in Figure 7.5 represents a fire safety decision-making process in an engineering context, where chance, decision, deterministic, and value nodes can be integrated. The relevant factors with roles can be summarized in Table 7.1.

The influence diagram represents decision-making for fire safety management in an industrial system, showing how uncertain events, human and system capabilities, and managerial decisions interact to determine safety, cost, and compliance outcomes. It integrates technical, human, and organizational factors into a single decision framework.

The influence diagram, as shown in Figure 7.5, illustrates a socio-technical decision-making structure for industrial fire safety, demonstrating how uncertainty, human factors, and managerial actions jointly shape safety outcomes. Probabilistic elements such as fire ignition and detection capability introduce inherent uncertainty into the system, while damage level captures the variable severity of consequences, ranging from acceptable to critical. These chance factors define the risk environment in which decisions are made and emphasize that fire events cannot be fully predicted or controlled, only mitigated through preparedness and system design.

Within this context, decision and enabling nodes play a pivotal role in determining final outcomes. Safety measures, decisions, and shutdown switchboard actions represent managerial and operational control, translating information into concrete responses that directly influence damage severity. Deterministic elements such as training, human–machine interface (HMI), and emergency warning systems enhance decision quality by improving situational awareness, communication, and response effectiveness. Together, these interactions culminate in the value node of safety, cost, and compliance, highlighting that effective fire safety management depends not only on technical detection systems but also on human competence and organizational decision-making, reinforcing the importance of integrated socio-technical approaches in industrial risk management.

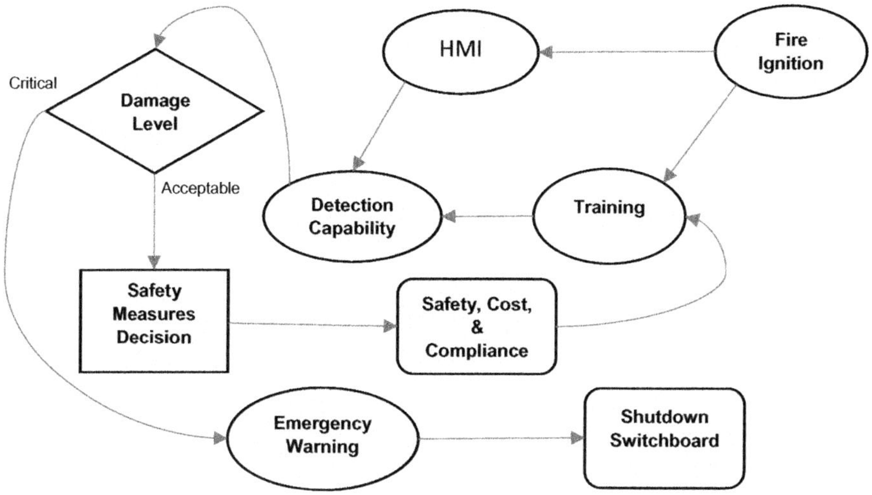

FIGURE 7.5 Influence Diagram for Fire Safety System.

Normative models do not aim to represent actual human behaviour but instead function as benchmarks for rational decision-making by offering optimal solutions under ideal conditions, which makes them valuable in policy design, engineering, and optimization, although their key limitation is that they ignore cognitive biases, emotional influences, information constraints, and uncertainty, leading real-world decision-makers to frequently violate their assumptions of perfect rationality.

7.2.2 Stochastic Model

A stochastic process describes the evolution of a sequence of random variables over time, representing the probabilistic counterpart to a deterministic process by incorporating inherent uncertainty, such that even with known initial conditions, it can produce multiple outcomes rather than a single predictable result.

The stochastic model includes utility as the back to its root to use probabilistic functions. Total utility value can be expressed as:

P = The probability of outcome
U = The Utility (or Value) of the specific outcome

$$T(U) = \Sigma\, P \times U$$

For example, consider the toss event in a cricket game where the toss coin is with two sides: head and tail, which means two outcomes. We can assume that win head with a 20% chance means P = 0.2 and the possibility of getting tail with an 80% chance means P = 0.8. If we consider the head as 100 and the tail as 0. Then, total utility can be calculated using the formula:

$$T(U) = (0.2\times 100)+(0.8\times 0) = 20$$

Therefore, the expected utility of the toss is 20, even though you will not win the toss.

7.2.2.1 Bayesian Theorem

Bayes' Theorem, named after the 18th-century British mathematician Thomas Bayes, is a mathematical formula for determining conditional probability. Conditional probability is the likelihood of an outcome occurring based on a previous outcome in similar circumstances.

Bayes' Theorem provides a mathematical framework for calculating conditional probability, which is the likelihood of an event occurring given that another related event has already occurred (Chickering et al., 2013). As a method of statistical inference, it enables the probability of a hypothesis to be updated by combining prior beliefs with new evidence, formally expressed as $P(A \mid B) = \frac{P(B|A)\,P(A)}{P(B)}$, where the posterior probability of event A is revised in light of observed evidence B.

Suppose we think of H as a hypothesis and E as evidence. Then the probability of H given E is as follows:

$$P(H|E) = \frac{P(E|H)\,P(H)}{P(E)}$$

$$= \frac{P(E|H)\,P(H)}{P(E|H)\,P(H)+P(E|\neg H)\,P(\neg H)}$$

$$= \frac{1}{1+\left(\frac{1}{P(H)}-1\right)\frac{P(E|\neg H)}{P(E|H)}}$$

For example, e-mail tracing from webmail can be described (as shown in Figure 7.6) following the Bayesian theorem. Assume that the word "sport" occurs in 70% of the spam messages in an account. Also, let us assume the "offer" occurs in 20% of desired e-mails. If 40% of the received e-mails are considered as a scam, and it receives a new message which contains "sport," what is the probability that it is spam?

Now, I assume that I have received 200 e-mails. The percentage of spam in the whole e-mail is 40. So, I have 80 spam e-mails and 120 desired e-mails out of 200 e-mails. The percentage of the word "sport" that occurs in spam e-mails is 70. It means 70% of 80 e-mails, and it makes 56. Now, it is calculated that 80 e-mails of 200 are spam and 56 of them include "sport" where 24 of them do not include "sport."

The question is what the probability of spam is where the mail includes the word "sport":

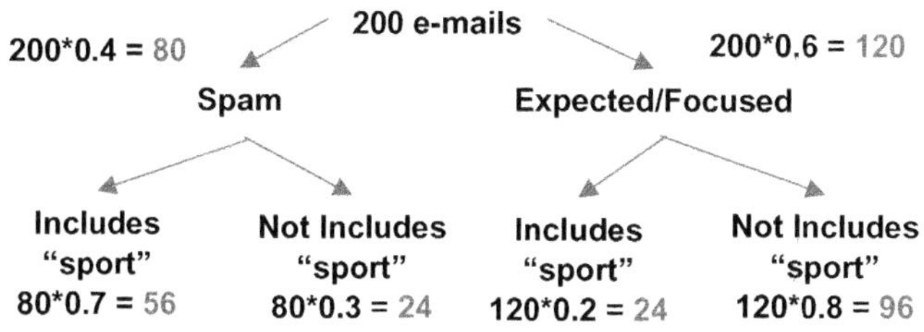

FIGURE 7.6 Probability Calculation of Tracing E-mails using Bayesian Theorem.

It needs to find the total number of e-mails which include "sport," that is, 56 + 24 = 80 mails include the word "sport." We can find the probability of spam if the mail includes "sport," that is, in 80 mails, 56 include "sport," which means **70% = 0.7** (probability)

The above solution can also be executed with **Bayes' Equation**:

A = Spam

B = Contains the word "sport"

P (spam I includes sport) = [P (includes sport I spam)×P (spam)]/P (includes sport)

P (includes sport I spam) = 0.7 (given in the question)

P(spam) = 0.4 (given in the question)

Now we can find the probability of an e-mail with the word "sport." We can compute that by adding "sport" to spam and desired e-mails. Such that:

P (includes sport) = 0.7×*0.4*+*0.2*×0.6 = 0.4

P (spam I includes sport) = (0.7×0.4)/0.4 = **0.70**

7.2.2.2 Markov Chain Model

A Markov chain model is a mathematical framework designed to support structured decision-making in dynamic systems (Feinberg, & Shwartz, 2012). It provides professionals with a disciplined way to determine which action should be taken at any given moment by focusing on the system's current condition and surrounding context. The model represents a system in terms of states, actions, and rewards or outcomes, and it uses these elements to determine how the system is likely to evolve over time. Depending on how states and decisions are defined, Markov models may be classified as finite or infinite, as well as discrete or continuous. At its core, the model relies on probabilistic dynamics to optimize decisions, making it particularly valuable in environments characterized by uncertainty rather than complete randomness.

What distinguishes the Markov chain model is its emphasis on the present. Decisions are made based solely on the current state of the system and the action taken in that state. The future state depends only on this immediate situation, not on the full sequence of past events that led there. This principle, known as the Markov property, removes the need to track the entire history of the system and allows decision-makers to focus on what matters most right now. As a result, Markov models are widely used in the design of intelligent agents and probabilistic planning systems, where the goal is to select actions that maximize long-term rewards while operating under uncertainty.

To illustrate, consider the operation of an information system that controls a machine or device. At any point in time, the system can observe the machine's current state, such as low energy, overheating, or normal operation. Based on this observation, it can choose an action that cools the machine, increases its speed, or shuts it down and receive a corresponding reward or outcome, such as energy savings, reduced downtime, or increased operational cost. Once the action is taken, the system transitions to a new state. Crucially, this transition depends only on the current state and the chosen action, not on how the machine reached that condition. This step-by-step evolution embodies the essence of the Markov model.

In practice, Markov chain models enable professionals to make optimal decisions in situations where outcomes are uncertain but not purely random. By continuously evaluating the current state and adjusting actions accordingly, the model supports learning and improvement over time, leading to better overall system performance. This adaptability makes Markov models especially suitable for complex, adaptive systems such as automated traffic control, predictive maintenance in smart factories, intelligent logistics networks, autonomous driving technologies, and energy optimization in smart buildings. In these contexts, the Markov approach provides a powerful mathematical foundation for building systems that respond intelligently to changing conditions and pursue long-term objectives with efficiency and resilience.

7.2.2.3 Markov Decision Processes

The evolution of the system in a Markov Decision Process (MDP) is governed by probabilistic rules. Once an action is taken, the system transitions to a new state according to a probability distribution that depends only on the current state and the chosen action. This principle is known as the Markov *assumption*. It asserts that the future is conditionally independent of the past: the next state is determined solely by where the system is now and what action is taken now, without any reliance on earlier states or decisions. This assumption simplifies modelling and computation while still capturing the essential uncertainty present in real-world systems.

The MDP represents a more advanced and structured extension of the Markov chain model, offering professionals a powerful analytical framework for decision-making in dynamic and uncertain environments. Within an MDP, a decision-making entity, often referred to as an agent, interacts with its environment over a sequence of time steps. At each time t, the agent observes the current state of the system, denoted asS_t, selects an action based on that observation, and then receives a reward R_t that reflects the immediate consequence or value of that action. This cycle of observation,

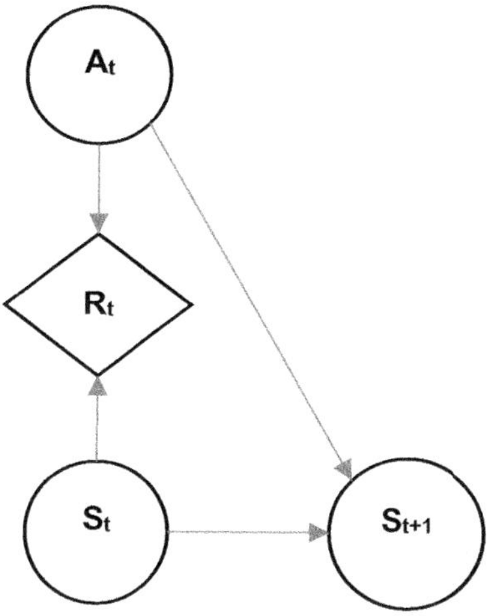

FIGURE 7.7 MDP in Decision-Making.

action, and reward, as shown in Figure 7.7, forms the foundation of rational decision-making in MDPs.

MDPs are often represented using decision networks, which provide a clear and systematic way to describe the relationships among states, actions, and rewards. In practice, particular attention is given to *stationary* MDPs. In a stationary MDP, the probabilistic laws governing the system remain constant over time. Specifically, the state transition probabilities $P(S_{t+1} \mid S_t, A_t)$ and the reward probabilities $P(R_t \mid S_t, A_t)$ do not change as time progresses. This stability allows the model to focus on long-term strategies rather than short-term fluctuations and enables efficient solution methods.

To support efficient representation and computation, stationary MDPs can be modelled using dynamic decision diagrams, which compactly encode the structure of the decision problem. Central to this representation is the state transition function, commonly denoted as $T(s' \mid s, a)$. This function specifies the likelihood that the system will move from a current state $s$to a subsequent state s' when a particular action a is taken. Together with the reward function, the transition function defines how decisions influence both immediate outcomes and future possibilities. Through this formal structure, MDPs provide a rigorous mathematical foundation for optimizing decisions over time in complex, uncertain environments.

7.2.2.4 MDP Components

An MDP is characterized by several core components that together explain how decisions are made under uncertainty. The first component is the set of *states* (S), which represent the possible conditions of the system. In a decision-theoretic framework, states function as outcomes that summarize all the information required for selecting appropriate actions.

The second component is the set of *actions* (A), defined as the possible acts an agent may take in each state. Actions allow the agent to influence the environment and shape future outcomes. Closely related to actions are the transition probabilities, denoted as $P(S_{t+1} \mid S_t, A_t)$. These probabilities describe the dynamics of the environment and capture the range of consequences that may result from taking a particular action in a given state.

Another essential element of an MDP is the *reward* function (R), which assigns value to states or state–action pairs. Unlike Bayesian approaches that focus primarily on belief updating, MDPs explicitly evaluate outcomes through rewards, thereby linking decisions to performance and objectives. In addition, MDPs incorporate a *discount factor* (γ), where$0 < \gamma < 1$, to reflect the relative importance of future rewards compared to immediate ones. Formally, the Markov property can be expressed as $P(S_{t+1} \mid S_0, \ldots, S_t, A_0, \ldots, A_t)$, which simplifies to dependence only on the current state and action. This means that the next state is determined by the present situation, independent of the system's past history.

Overall, an MDP consists of an agent, a set of states, available actions, transition dynamics, rewards, and an optimal policy. The agent interacts with an environment that can exist in many different states, and the MDP framework specifies how actions influence transitions between those states. The central decision problem is to identify, for each state, the action that maximizes the expected reward. A policy provides this guidance by mapping each state to the action the agent should take, while rewards are generated based on the current state and chosen action.

In practical terms, defining the state space is often the most challenging task, as a state may include information about position, resources, and environmental conditions. Rewards are tied to state quality, with more desirable states yielding higher rewards. To execute an MDP simulation, the states, actions, probability distributions, and rewards must be clearly specified. At each time step, the current state is observed, and an action is selected accordingly. The process typically begins from an initial state, which is often chosen at random from the set of states, and then evolves step by step as decisions are made.

For example: The early warning system in a chemical industry can be described with a Markov network as shown in Figure 7.8. The warning system checks the level of safety and triggers the alarm to build awareness in the plant.

The Markov value can be calculated at each event using the Markov network equation as follows:

$$\text{Value} = V(S_1) = R(S_1) + \gamma \text{ max\_A } \Sigma\text{\_S' } P(S' \mid S_1, A)\, V(S')$$

Considering the above equation, we can calculate the value for the event "SS= Shutdown Switch" as below:

$$\begin{aligned} V_{ss}(S) &= -3 + 0.70 \times \{(-3 \times 0.3) + (-7 \times 0.1)\} \\ &= -10 + 0.70 \times (-16) = -3 - 1.12 = -4.12 \end{aligned}$$

The values for all events, considering probabilities and rewards, can be presented in Table 7.2.

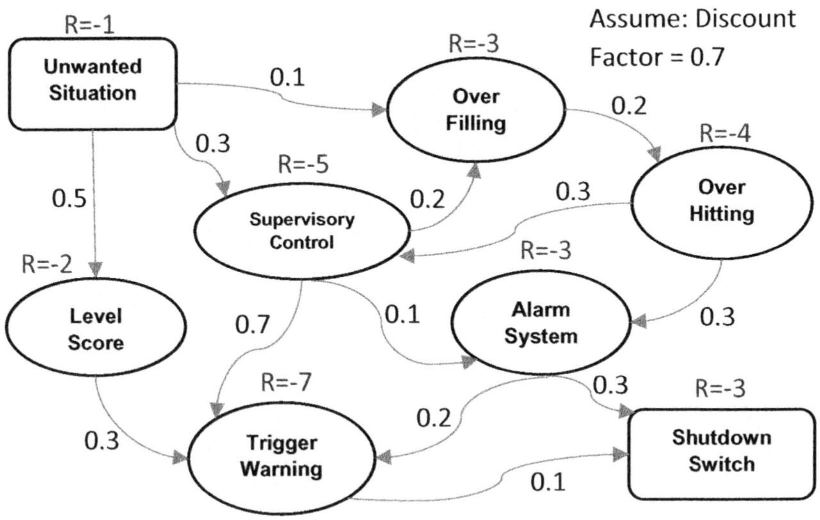

FIGURE 7.8 Markov Network for an Early Warning System.

The detailed calculations for all events are described below:

$$V\ (\text{Unwanted Situation}) = -1 + 0.7\,(-3 - 0.1 + -.5 - 0.3 + -2 - 0.5) = -2.96$$

$$V\ (\text{Over Filling}) = -3 + 0.7\,(-4 - 0.2) = -3.56$$

$$V\ (\text{Over Hitting}) = -4 + 0.7\,(-5 - 0.3 + -.3 - 0.3) = -5.68$$

$$V\ (\text{Supervisory Control}) = -5 + 0.7\,(-3 - 0.2 + -3 - 0.1 + -7 - 0.7) = \mathbf{-9.06}$$

$$V\ (\text{Level Score}) = -2 + 0.7\,(-7 - 0.3) = -3.47$$

$$V\ (\text{Alarm System}) = -3 + 0.7\,(-7 - 0.2 + -3 - 0.3) = -4.61$$

$$V\ (\text{Trigger Warning}) = -7 + 0.7\,(-3 - 0.1) = \mathbf{-7.21}$$

From the above probability transition matrix, it is evident that professionals need to give proper attention to supervisory control and trigger warning as critical events in the warning system operation.

7.2.3 Qualitative Model

The qualitative model is depicted by the collection and analysis of qualitative data (observations, interviews, focus groups, surveys, conversational analysis, ethnographies, and secondary research. Qualitative methods are a suitable approach to extract valuable insights by analysing data. This model is rigorous and easier with

TABLE 7.2
Probability Transition Matrix in a Markov Chain Model for Warning System

Reward	Probabilities	Unwanted Situation	Over Filling	Over Hitting	Supervisory Control	Level Score	Alarm System	Trigger Warning	Shutdown Switch	Value = $V(s_1) = R(s_1)+\gamma$ max_a Σ_s′ $P(s' \mid s_1, a)\, V(s')$
–1	Unwanted Situation	0.1	0.1		0.3	0.5				**–2.96**
–3	Over Filling		0.1	0.2						**–3.56**
-4	Over Hitting				0.3		0.3			**–5.68**
–5	Supervisory Control		0.2		0.1		0.1	0.7		**–9.06**
–2	Level Score					0.1		0.3		**–3.47**
–3	Alarm System						0.1	0.2	0.3	**–4.61**
–7	Trigger Warning							0.1	0.1	**–7.21**
–3	Shutdown Switch								0.1	**–4.21**

its own set of data to ideate new knowledge, particularly to explore valuable insights through qualitative findings.

This model involves the delicate planning of a research approach that can collect participants' viewpoints. According to qualitative studies, the research focus emphasizes to analyse data derived from respondents and their perceived experiences and opinions. Professionals can use NVivo or other qualitative tools to analyse qualitative data.

Secondary data is one of the most popular qualitative methods to extend the analytical process, particularly in the marketing research industry. Currently, professionals apply secondary research as a qualitative approach for data mining, especially for business process intelligence. The data mining can be executed through predictive and descriptive analytics. Marketing professionals can extend brand equity research to identify product perceptions from the market. For example, we can analyse the secondary data and extend qualitative data analysis to extract information regarding the cheese market in Australia.

Bega Cheese Ltd is an Australian publicly listed dairy and food company with origins as a farmer cooperative founded in 1899 in the Bega Valley, New South Wales (Bega Group, 2024). Over time, it has transformed from a commodity-focused dairy producer into a diversified branded food business, shifting its emphasis toward value-added consumer brands. Today, Bega's portfolio extends beyond cheese to include spreads, nutritional powders, infant formula, peanut butter, and salad dressings, positioning it as "The Great Australian Food Company". In FY25, the company reported normalized EBITDA of $202.0 million, a 23% increase on the prior year, reflecting improved market share and margin momentum despite cost pressures (Bega Group Limited, 2025).

Within Australia's cheese market, Bega is a leading domestic brand, particularly in cheddar and processed cheese but faces intense competition from Fonterra's Perfect Italiano, which benefits from global scale and supply chain efficiencies. Bega's competitive advantage lies in its strong Australian heritage and brand identity, while its long-term competitiveness depends on continued efficiency, innovation, and strategic positioning as it transitions further into branded, value-added products.

Bega's growth has been accelerated through major acquisitions, notably Vegemite in 2017 and Lion Dairy & Drinks in 2021, which added iconic brands such as Dairy Farmers and Zooper Dooper. These moves strengthened its domestic market position and broadened its operations beyond dairy. According to its 2024 Annual Report, Bega generated approximately A$3.52 billion in revenue, underscoring its diversification into wider food and beverage categories (Bega Group, 2024).

Despite facing competition from global dairy firms, supermarket private labels, and shifting consumer preferences toward premium and artisan cheeses, Bega maintains leadership through a dual strategy of offering both affordable everyday products and premium options. Recognized as Australia's leading cheese brand in 2020, Bega demonstrates strong adaptability and brand equity. Continued investment in innovation and quality will remain critical as competition intensifies, particularly between private labels and premium rivals.

7.2.4 Quantitative Model

Engineering professionals seek to identify how variables or attributes are correlated and how they influence numerical outcomes through causal effects. Quantitative methods are well suited to this approach, as they enable the development of hypotheses, justify the selection of variables, and provide answers to the research questions. The hypothesis is verified and validated through analytical methods by analysing data (i.e., surveys, experiments, controlled observations, interviews, or longitudinal studies) and extending statistical analysis.

The main objective of a quantitative model is to support professionals or project teams to get findings through an extensive analytical process. Data collection method would be critical before the analytical process. Experiment, survey, computer simulation, controlled observation, survey questionnaire, longitudinal studies, and interview are popular quantitative data collection methods. Then, data sampling techniques are used to choose an appropriate sampling process during the data collection from the entire population. The quantitative model extracts results that can be interpreted through probabilistic and statistical calculations and then, it can be replicated and applied to the entire population. Professionals use SPSS (Statistical Package for the Social Sciences), SmartPLS, or other suitable quantitative tools to analyse collected data that can extract relationships between variables and other statistical values.

For example: Professionals can develop an online questionnaire survey to assess brand equity for cheese products in Australia using Qualtrics and create questionnaires for perceptions of quality, brand familiarity, brand equity, and socio-demographic variables. Quantitative data can be collected from respondents and cleaned. Then, they can extend data analysis using the SPSS tool. Initially, professionals can execute descriptive statistics. They can also execute independent sample t-test (comparing two group means), paired/matched sample t-test (comparing two individual means), and predictive models through linear and logistic regression.

7.3 DISCUSSION AND SUMMARY

In summary, this chapter has examined a range of decision-making tools that support engineering professionals in navigating uncertainty within complex organizational environments. While deterministic decisions assume certainty of outcomes, most real-world engineering challenges are inherently stochastic, requiring the evaluation of probabilities and imperfect information. Decision-making tools serve as structured and intuitive representations of these situations, enabling professionals to articulate assumptions about uncertainty, assess preferences among alternatives, and integrate relevant contextual knowledge. By combining normative and stochastic modelling with qualitative and quantitative approaches, these tools accommodate experience, emotional influences, and situational awareness, thereby strengthening the foundation for reasoned and defensible decisions.

Furthermore, the chapter has distinguished between normative models, which prescribe optimal decision rules based on logic, mathematics, and economic rationality, and stochastic models, which explicitly address uncertainty and dynamic conditions. Techniques such as brainstorming, SWOT analysis, decision trees, and influence diagrams illustrate normative approaches that assume rational actors and complete information. In contrast, Bayesian reasoning and MDPs offer more advanced and adaptive frameworks for decision-making under uncertainty, capturing probabilistic relationships and evolving system states. Together, these models demonstrate how structured decision tools can transform complexity into clarity, enabling engineers to make informed, rational choices even in unpredictable and high-risk contexts.

BIBLIOGRAPHY

Bega Group Limited. (2024). *Annual report 2024*. https://begagroup.com.au/investors/annual-reports/

Bega Group Limited. (2025). *FY25 full year results announcement*. Australian Securities Exchange. https://www.asx.com.au.

Boucherie, R. J., & van Dijk, N. M. (Eds.). (2017). *Markov decision processes in practice* (1st ed.). Springer International Publishing. https://doi.org/10.1007/978-3-319-47766-4.

Carmody-Bubb, M. (2023). *Cognition and decision making in complex adaptive systems: The human factor in organizational performance*. Springer Nature.

Chickering, D. M., Heckerman, D., & Meek, C. (2013). A Bayesian approach to learning Bayesian networks with local structure. https://doi.org/10.48550/arxiv.1302.1528.

Feinberg, E. A., & Shwartz, A. (Eds.). (2012). *Handbook of Markov decision processes: Methods and applications* (Vol. 40). Springer Science & Business Media.

Fenton, N. E., Neil, M., & Taylor & Francis. (2019). *Risk assessment and decision analysis with bayesian networks* (2nd ed.). CRC Press.

Gal, Y., & Pfeffer, A. (2008). Networks of influence diagrams: A formalism for representing agents' beliefs and decision-making processes. *The Journal of Artificial Intelligence Research*, 33, 109–147. https://doi.org/10.1613/jair.2503.

GeNIeModeler. *Programmer's manual, version 4.1.R0, Built on 9/15/2023*. BayesFusion, LLC.

Jamieson, G. A., & Skraaning, G. (2018). Levels of automation in human factors models for automation design: Why we might consider throwing the baby out with the bathwater. *Journal of Cognitive Engineering and Decision Making*, 12(1), 42–49. https://doi.org/10.1177/1555343417732856.

Kochenderfer, M. J. (2015). *Decision making under uncertainty: Theory and application*. MIT Press.

Phadermrod, B., Crowder, R. M., & Wills, G. B. (2019). Importance-performance analysis based SWOT analysis. *International Journal of Information Management*, 44, 194–203. https://doi.org/10.1016/j.ijinfomgt.2016.03.009.

Thomas, L. C. (1995). Markov decision processes: Markov decision processes: Discrete stochastic dynamic programming. *Journal of the Operational Research Society*, 46(6), 792–793. https://doi.org/10.1057/jors.1995.110.

U.S. Chemical Safety and Hazard Investigation Board (CSB). (2017). *Airgas Facility Fatal Explosion Investigation Report (Report No. 2016-04-I-FL)*. U.S. Chemical Safety and Hazard Investigation Board. www.csb.gov/assets/1/20/Airgas_Investigation_Report_-_Final_-_2017-02-28.pdf

8 Theory and Model Applications

Bayesian and Markov Network

8.1 INTRODUCTION

Judgement and decision-making (JDM) is an important discipline in engineering. This is because technical outcomes depend on a combination of calculation and how individuals and organizations manage risk and respond under pressure. Real-world engineering environments often involve incomplete information, commercial pressure, and time-bound decisions. These conditions lead to uncertainties, especially vulnerable to cognitive biases, heuristic shortcuts, and conflicting interests that demand model-driven decision-making. Intuitive judgement is particularly prone to bias under such time pressure and reputational concerns.

The model-driven decision-making emphasizes the internal mental processes involved in decision-making, including perception, reasoning, intuition, memory, experience, and emotional responses. In practice, individuals often depend on these cognitive processes rather than strictly logical or data-driven approaches, particularly in dynamic situations. This model is especially relevant when decisions must be made under time constraints, stress, limited information, or strong psychological influences. In high-risk contexts such as medical emergencies or aviation operations, cognition-driven decision-making helps explain how even highly experienced professionals may commit errors despite their expertise. For these complex planning scenarios, a model-driven approach is appropriate to bring rigour and transparency to analyse.

Decision-making has long been recognized as one of the most critical yet imperfect aspects of human behaviour. Even in structured environments, choices are shaped by bounded rationality, cognitive biases, uncertainty, and emotional states. Traditional decision theory assumes that decision-makers are "rational actors" who have complete access to information and can compute outcomes perfectly. They assess all viable options, calculate expected results, and choose the one that yields the best outcome. In real life, organizations seldom function under these ideal conditions. Information is often incomplete, time is limited, resources are scarce, and preferences can be inconsistent or unclear, which creates uncertainty.

Effective decision-making under uncertainty is central to complex adaptive systems, where seconds often determine whether an event becomes a near miss or a catastrophe. Engineering projects under high-risk conditions require rational decision-making processes that combine logical evaluation with human mental processing

DOI: 10.1201/9781003685449-8

restrictions and probabilistic calculations. People under time constraints and unclear conditions and organizational performance goals tend to use established theory and model applications that can avoid mental shortcuts, which produce systematic errors.

Decision-making theories and applied models enable the evaluation of critical situations within complex environments, thereby supporting the achievement of optimal outcomes. By treating each state as an independent decision process, these models emphasize strengthening the internal validity of individual systems and assessing how effectively each function performs based on the information and structural constraints available to it, rather than interpreting decisions as simple causal links within a linear sequence.

Decision-making models are designed to manage complexity and support professionals in reaching rational and defensible decisions. In contrast to cognition-driven models, which emphasize human biases and heuristics, model-driven approaches are grounded in formal decision-making theories such as probability theory and decision theory, explicitly integrating uncertainty into the analytical process. This chapter examines Bayesian networks (BNs) and Markov chain models in detail, using multiple examples to interpret complex decision-making scenarios. Through these explanations, professionals gain clearer insight into the interactions among key variables in critical situations and are better equipped to derive informed conclusions and select appropriate decision-making strategies.

Decision-making models provide structured approaches for addressing complex situations characterized by uncertainty. A range of use cases is employed as illustrative examples to show how systematic modelling through model-driven approaches such as BNs and Markov chains can contribute to safer outcomes in critical incidents. These frameworks support the rigorous evaluation of complex situations and available alternatives, demonstrating how Bayesian and Markov-based applications enable informed and defensible decision-making in uncertain environments.

8.2 THEORY AND MODEL APPLICATION

Uncertainty is a critical factor in tackling critical situations and probability plays a vital role in making decisions. Expected utility theory is one of the fundamental decision-making theories to explain the outcomes of activities. The expected utility can be calculated by multiplying the utility of each event by the probability of occurrence and then summing up the results of all events. Professionals apply subjective probabilities when the outcomes are uncertain and critical to making decisions. The degree of risk aversion is a critical factor in deciding how professionals respond to uncertain situations across the utility function.

Markov decision processes (MDPs) extend Markov chains by including actions and rewards and can be used to assess the best decision policies in the presence of uncertainty. The following is a refined MDP model, drawing upon empirical uses in the literature, including Markov-switching models to predict financial stress and recession. The model considers the crisis as a discrete-time state transition system with the states indicating economic states, actions, policy choices, and rewards measuring the results of the crisis like GDP effects or financial stability.

The expected value of an action A is defined based on the weighting of the preferences W(C) for the consequences C, the probability of occurrence P, and the expected value of the action defined as:

$$EW(A) = \Sigma P \times W(C)$$

A normal distribution (also called a Gaussian distribution) is a bell-shaped curve that presents how values of a continuous variable are spread out, with most values close to the average and fewer extreme values. A Gaussian distribution (also called a normal distribution) is a common way to describe values that cluster around an average (like people's heights or test scores). However, it does not work well in cases where values cannot be negative or too extreme, for example, when measuring altitude. However, Bayesian models can link to a normal distribution and apply probabilistic calculations to make decisions and update beliefs based on added information.

Another important theoretical approach is to understand conditional probability that can measure the probability of an event occurring, given that another event has already occurred. It provides a way to update probabilities based on the latest information. Conditional probability helps professionals to update their beliefs based on current information in an uncertain situation.

Comparing the two models highlights how psychological and analytical dimensions interact. The cognitively driven model reveals obstacles that prevent people from following rational procedures, while the model-driven system provides a measurable benchmark for rationality. One explains failure; the other defines success. When applied together, they form a comprehensive decision-making framework in which human awareness and structured reasoning mutually reinforce each other.

The model demonstrates that even with available data, poor perceptions and excessive group pressure can lead to delayed and inconsistent action. The model also suggests that decision quality improves when social hierarchies and political interference are minimized. The model emphasizes the need to strengthen psychological preparedness, promote critical discussion, and design communication structures that encourage open expression of dissenting opinions in high-risk environments.

8.2.1 Bayesian Network

BN was constructed to model the probabilistic relationships between key safety variables and incident outcomes. The structure reflects both theoretical reasoning and case-specific evidence, allowing for scenario simulation and risk prioritization. A BN requires identification of variables that could influence the final outcome.

A BN is the modelling of systems in which several uncertain factors interact simultaneously. Instead of having a single likelihood update, a BN incorporates relationships among design, construction quality, load contribution, damage indicators and professional assessments, as well as their effect on the final result. This structure makes it ideal for structural crisis evaluation because sudden structural failure is rarely due to one factor in isolation. Instead of relying on past experience or insufficient information, a Bayesian approach integrates prior knowledge with new data to produce a mathematically based probability that reflects the most current situation. The BN is organized in a linear structure that illustrates how design factors and observable structural symptoms combine to influence crisis response decisions.

BNs provide a normative framework for decision-making under uncertainty by representing probabilistic dependencies among variables and enabling systematic belief revision as new evidence emerges. The framework applies directly to modelling how decision-makers should update probability estimates when confronted with contradictory information. This addresses Decision 2's core failure – the underweighting of diagnostic evidence from Swedish detection relative to entrenched "controlled-incident" priors. The model structure maps directly onto the Soviet leadership's decision environment: uncertain radiation severity, observable detection evidence, a binary evacuation choice, and competing utilities balancing public health against political cost.

Usually, professionals apply BNs to estimate probabilities for real-world applications that follow a normal distribution such as temperature control and sensor-based digital applications. BNs use mathematical derivations of normal distributions to calculate probabilities when making predictions that support professionals to make decisions. These networks can instantly project predictions by updating probabilities when new data reach in uneven situations.

The Bayesian theorem establishes a philosophical stance that enhances professionals' thinking processes to address uncertainty by applying probabilities. It recommends that probabilities denote belief instead of just considering previous events. Bayesian probability can measure how much professionals have faith in something that will happen in the next consequences.

The BN is capable of updating beliefs with the latest information as new data arrives, and it can adjust the pre-existing probabilities to reinforce the decisions. This network applies a reasoning process, both addressing to predict future events (forward reasoning) and figuring out causes based on outcomes (backward reasoning). Briefly, BN enriches professionals' thinking process and enables them to make better decisions by constantly updating existing knowledge through contextual information.

The BN is a graph-based representation of probability distributions for an event. Nodes in the network indicate random variables, and edges designate probabilistic relationships. Each node has a conditional distribution, which illustrates the probability of that variable given its parents in the network.

There are two phases to build a BN: the first phase determines a model's graphical structure that includes two stages: identifying the variables in a situation that should be included in the model and defining the state of each variable. The second stage requires establishing the relationships between the selected variables and determining the direction of those relationships. After defining the graphical structure, the next phase is to select the parameters for each variable, which represent the strength of the relationships in the BN structure.

Focusing on the rational support in decisions and avoiding crisis, it is indicated through the refinement of the BN to the given diagram and through the usage of decision, behavioural, and result nodes. BNs are commonly used to make explicit decisions in uncertain situations, and professionals can apply utility functions to get optimal options in solving practical problems. BNs incorporate Influence diagrams to make rational decisions, following utilities, decisions, and chance nodes. It provides a robust decision-making framework that can optimize decisions under uncertainty.

The BN supports professionals to assess available alternatives and can plan to choose the best option in an uncertain situation. This is an intuitive, graphical representation of a decision-making process that illustrates the identified problems and alternatives. The maximum expected utility of an optimal strategy can be computed automatically. They represent causal relations through a graph, strength of relations through probabilities, preferences through utilities, and recommendations through maximising expected utility. The tasks suitable for normative systems can be divided into three subclasses: forecasting, interpretation, and planning.

The systematic support of high-risk engineering project decisions becomes possible through official probabilistic models like the BN model. The structured frameworks of the model-driven approach enable professionals to represent uncertainty together with causal links and outcome probabilities. The complex offshore drilling environment benefits from BNs because these models manage multiple interacting safety barriers effectively. The following section presents a BN analysis of the Volkswagen emissions scandal.

The process of updating beliefs through new evidence depends on Bayes' rule, which serves as the basis for probabilistic decision-making. The mathematical structure of BNs depends on Bayes' rule to merge prior information with observed data for determining posterior probability values.

The mathematical formula for Bayes' rule appears as follows:

$$P(Xi \mid E)=P(E \mid Xi)\ .P(Xi)P(E)$$

where $P(Xi \mid E)$: posterior probability of event Xi given evidence E $P(E \mid Xi)$: likelihood of evidence E if event Xi is true $P(Xi)$: prior probability of Xi $P(E)$: marginal probability of evidence E.

The BN uses nodes to represent random variables like cement quality and blowout preventer reliability, while arcs show the relationships between them. The conditional probability tables (CPTs) within the system show how parent variables affect the outcomes of child variables. The BN system updates all probability values throughout the network after users input evidence such as a failed pressure test.

This explains how BNs use Bayes' rule to update probabilities based on added information. Bayes' rule formula can be mentioned as the equation:

$$P\left(Y \middle| X\right) = \frac{P\left(X \middle| Y\right) \cdot P(Y)}{P(X)}$$

where this helps to update the probability of Y happening, given that X has occurred.

Bayes' network follows causal relationships through causal/series, intercausal/convergence, and divergence. The relationships and relevant equations can be illustrated in Figure 8.1.

The BN is constructed entirely from observable and documentable factors to ensure that it can be quantified and implemented in GeNIe. The GeNIe software allows users

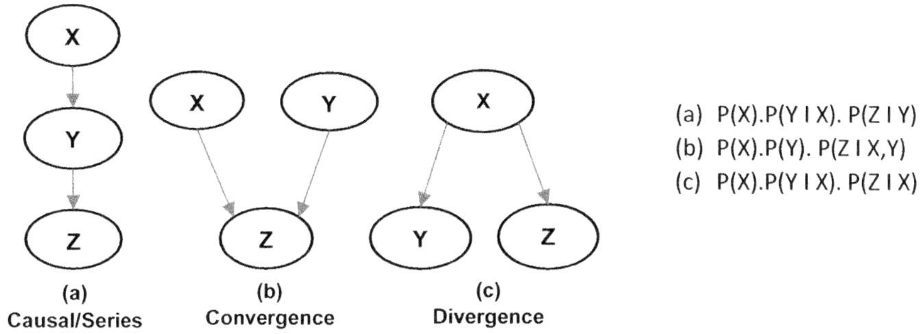

FIGURE 8.1 Notations and Equations for Causal Relationships in BN.

to analyse different scenarios through this network, demonstrating how probabilistic reasoning systems could have delivered safer decision options. Implementation process in GeNIe can be described as follows:

- All nodes and arcs are created according to the structure above.
- Chance-node CPTs are entered numerically; utility node values are set in value mode (not %).
- The model is evaluated under two conditions: Base case (no evidence) to observe priors and anomaly scenario (evidence entered).
- GeNIe automatically propagates posterior probabilities and calculated expected utilities.

For example, we can apply BN to explain the probabilities of having train disruption considering two variables, such as communication signal loss and severe weather. Researchers extend quantitative research to determine the probability of one event. In this example, we can easily get the probabilities of having train disruptions by analysing secondary data that is available related to communication signal loss and severe weather. Then, the true percentage can be calculated to form CPT. Initially, the probability of the true condition can be calculated through research outcomes. We can assume that the probability of a true case in having communication signal loss and severe weather are P(CS) = 0.2 and P(BW) = 0.4. The CPT (as shown in Figure 8.2) and the BN can be drawn as follows:

The above BN and probabilistic matrix, considering past experience, can be applied in predictive digital applications to aware passengers. Professionals can accommodate this probabilistic calculation in the machine learning platform for experience-driven and decision-making platforms.

Professionals can assess decision-making situations, identify appropriate cognitive theories, and apply a model-driven approach like BN to demonstrate the requirements and make a rational decision with practical implications. When professionals apply probabilistic calculations based on recent knowledge using Bayesian theory, it provides a logical and judgemental decision-making for an organization, such as to develop an information system or to operate cybersecurity applications.

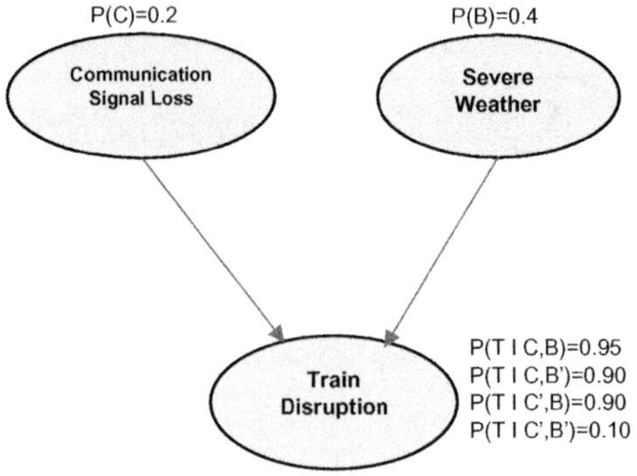

Conditional Probability Table

C	B	P(T=T I C,B)	P(T=F I C,B)
T	T	0.95	0.05
T	F	0.90	0.10
F	T	0.90	0.10
F	F	0.10	0.90

FIGURE 8.2 BN for Train Communication System.

8.2.1.1 Use Case Scenario

In an era of increasing environmental awareness and regulatory enforcement, decision-making within organizations has become a multidimensional challenge. This chapter explains a decision-making situation on the Volkswagen emissions scandal, commonly known as "Diesel gate," to analyse and improve corporate decision-making processes. Specifically, it applies BN as a decision-making model to two critical moments in the scandal to evaluate the role of human cognition and structured methodologies in ethical lapses and corporate crisis management.

In 2015, the United States Environmental Protection Agency (EPA) revealed that Volkswagen had deliberately equipped over 11 million diesel vehicles worldwide with "defeat devices" designed to manipulate emissions readings during regulatory testing. While these vehicles appeared compliant under controlled laboratory conditions, they emitted pollutants up to 40 times the legal limit during real-world operation. This deliberate deception was a strategic attempt to expand market share without committing to the development of cleaner, compliant technologies, resulting in one of the most prominent corporate ethics scandals of the modern era.

Senior leadership and engineering personnel within Volkswagen were directly implicated in both the initial implementation of the defeat devices and the subsequent delay in addressing the scandal. These actions significantly exacerbated the company's legal liabilities and reputational damage. The case presents a compelling context in which to examine the influence of psychological biases, flawed judgement, and inadequate application of decision-making frameworks in producing far-reaching organizational failures.

The Volkswagen emissions scandal drew global attention when the U.S. EPA found the company had deliberately violated the Clean Air Act by installing illegal "defeat devices" in diesel engines. These systems manipulated emissions during lab testing, while actual emissions exceeded legal limits by up to 40 times.

This deliberate regulatory breach triggered widespread legal actions across the United States, Europe, Canada, and Asia. Volkswagen faced over USD 30 billion in fines, settlements, and recalls. In the United States, the case led to felony charges and arrests of senior executives. The scandal remains a pivotal example of corporate misconduct and the importance of environmental compliance.

The Volkswagen emissions scandal encapsulates three critical decision-making junctures that collectively defined the scope and severity of the crisis. The first involved the strategic decision to install defeat devices, motivated by competitive market pressures and regulatory constraints, which reflected a deliberate prioritization of commercial interests over ethical and legal obligations. The second scenario concerned the active concealment of the manipulation, demonstrating a prolonged organizational commitment to deceptive practices despite escalating legal and reputational risks. The third scenario emerged following regulatory exposure, wherein Volkswagen was confronted with a pivotal choice: to adopt a posture of transparency and cooperation or to engage in denial and deflection. Each of these decisions exerted considerable influence over the company's legal liabilities, public trust, and long-term strategic trajectory.

A BN is well-suited for addressing complex, high-stakes decision-making scenarios like Volkswagen's post-Diesel gate response. BNs are probabilistic models that represent variables and their conditional dependencies, enabling structured reasoning under uncertainty.

This approach allows Volkswagen to simulate various response strategies ranging from full cooperation to denial and assess their impact on regulatory penalties, public perception, stock performance, and brand recovery. By mapping causal relationships, the model provides a transparent framework for evaluating trade-offs and optimizing outcomes.

BNs have been successfully used in modelling crisis chains and decision complexity, making them ideal for Volkswagen's shift from reactive crisis response to proactive, strategic planning. This supports timely, evidence-based decisions that rebuild trust and reduce legal and reputational risks.

The BN developed for Volkswagen's post-Diesel gate strategic decision-making models the interdependencies among key variables, including regulatory cooperation, public communication, penalty risk, market reaction, and profitability outcomes. This probabilistic framework enables decision-makers to systematically forecast the impact of alternative response strategies under conditions of uncertainty.

The primary components of the model include:

- Regulatory cooperation: This input node represents the extent of Volkswagen's collaboration with regulatory authorities following the scandal. It is defined by two states: high and low. Greater cooperation is associated with a reduced likelihood of severe regulatory penalties.
- Public communication: This input node captures the clarity, transparency, and effectiveness of Volkswagen's communication with the public and stakeholders. It also includes two states: good and poor. Effective public communication increases the probability of a positive market response.
- Regulatory penalty risk: An intermediate node influenced by the level of regulatory cooperation. It estimates the likelihood of Volkswagen facing high or low regulatory penalties based on its demonstrated willingness to cooperate.

- Market dynamics impact: Another intermediate node, this reflects how the market is likely to respond to Volkswagen's crisis communication efforts. It accounts for either a positive or negative market reaction depending on the transparency and timing of the messaging strategy.
- Profit recovery: This is the final outcome node, determined by both regulatory penalty risk and market dynamics impact. It forecasts whether Volkswagen's recovery trajectory will be fast or slow, based on the combined influence of legal and reputational outcomes.

The model structure links decision variables (cooperation and communication) with intermediate outcomes (penalty risk and market impact), leading to a long-term business result (profit recovery). Each node is governed by a conditional probability distribution that updates dynamically based on changes in the input variables. This allows Volkswagen to simulate a range of strategic scenarios and identify optimal pathways for rebuilding profitability and trust.

The CPTs presented in Tables 8.1–8.3 define the probability distributions for each node in the BN. These values are based on logical assumptions aligned with the Volkswagen post-Diesel gate decision context.

The model assumes that Volkswagen's recovery trajectory is primarily influenced by the level of cooperation with regulatory authorities and the transparency of its public communication strategy. It presumes that these factors have a direct and measurable effect on regulatory responses, market perceptions, and profit outcomes. Additionally, the model assumes that the conditional probability values defined for each node, although subjectively estimated, are a reasonable representation of real-world outcomes under each scenario. It also assumes a static decision context in

TABLE 8.1
Node Properties for Regulatory Cooperation, Public Communication, and Regulatory Penalty Risk

Node/Event	Definition	Value
Regulatory Cooperation	High	0.6
	Low	0.4
Public Communication	Good	0.7
	Poor	0.3
Regulatory Penalty Risk	High	0.7
	Low	0.3

TABLE 8.2
Node Properties for Market Dynamics Impact

Public Communication	Good	Poor
Positive	0.8	0.3
Negative	0.2	0.7

TABLE 8.3
Node Properties for Profit Recovery

Regulatory Penalty Risk	High		Low	
Market Dynamics Impact	**Positive**	**Negative**	**Positive**	**Negative**
Fast	0.4	0.2	0.9	0.5
Slow	0.6	0.8	0.1	0.5

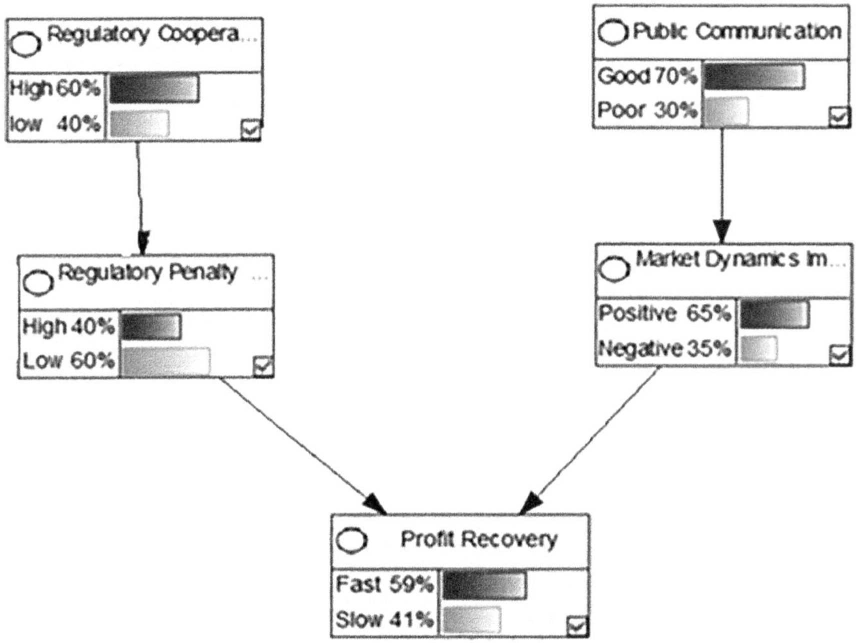

FIGURE 8.3 Bayesian Model – Baseline Scenario (Screenshot from GeNIe).

which no external shocks (e.g., geopolitical crises or supply chain failures) interfere with the modelled relationships.

GeNIe Modeler was used to perform the analysis, yielding the following results under varying strategic scenarios. In the baseline scenario (no evidence set), the model estimated a 59% probability of fast recovery and 41% for slow recovery, reflecting neutral assumptions across all decision variables.

The Bayesian Decision Network was evaluated using evidence representing the operational and organizational conditions that existed during the 2022 Titan dive season, when OceanGate continued missions despite emerging warning signs. Each evidence variable corresponds to a state that was either reported directly in official investigations or reliably described in reputable media coverage. Entering this evidence into *GeNIe* triggered posterior updates that produced the model outputs shown in Figure 8.3.

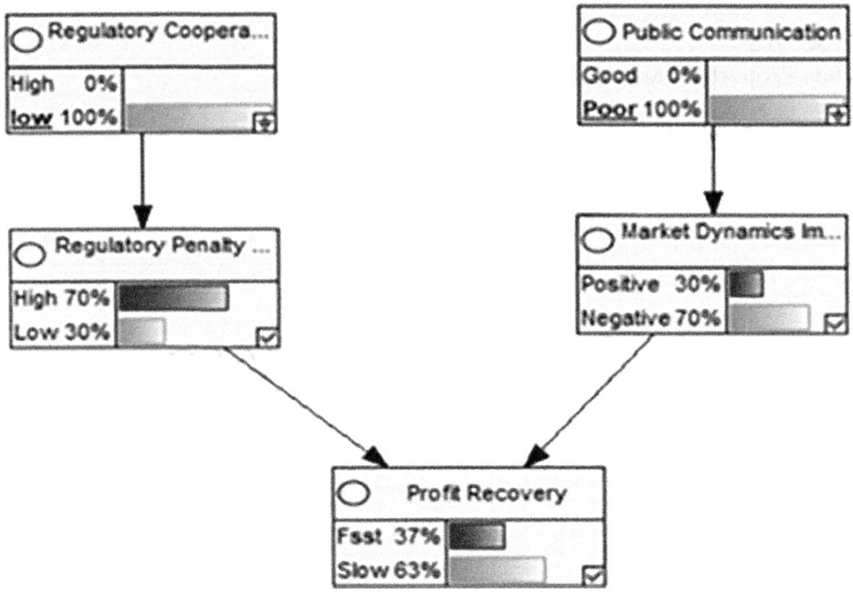

FIGURE 8.4 Bayesian Model – Worst-Case Scenario.

In the worst-case scenario, where low regulatory cooperation and poor public communication were selected as evidence, the probability of slow recovery increased to 63%, while fast recovery declined to 37%. This scenario highlights how non-cooperative regulatory behaviour and poor transparency amplify legal and reputational risks, significantly delaying recovery.

Conversely, the best-case scenario, defined by high cooperation and good communication, resulted in the most favourable outlook. Here, the model projected a 73% chance of fast recovery and only 27% for slow recovery, underscoring the importance of proactive stakeholder engagement and regulatory compliance.

These outcomes illustrate how strategic decisions drive recovery probabilities. The model supports informed, risk-sensitive decision-making by showing that openness and cooperation produce measurable benefits for long-term business performance and reputational resilience.

The BN model was selected for this analysis due to its strength in handling uncertainty, interdependent variables, and incomplete information, features highly relevant to Volkswagen's post-Diesel gate strategic decision environment. Unlike deterministic models, BNs offer a probabilistic approach that enables scenario forecasting and evidence-based decision-making under risk, making it ideal for complex corporate crises where outcomes are not easily predictable.

In the case of Volkswagen, decision-making involved balancing legal compliance, reputational repair, and financial performance, all under dynamic regulatory and market pressures. A Bayesian model provides the structure to simulate how variables such as regulatory cooperation and public communication influence penalty risk, market response, and profit recovery. This framework allows decision-makers to

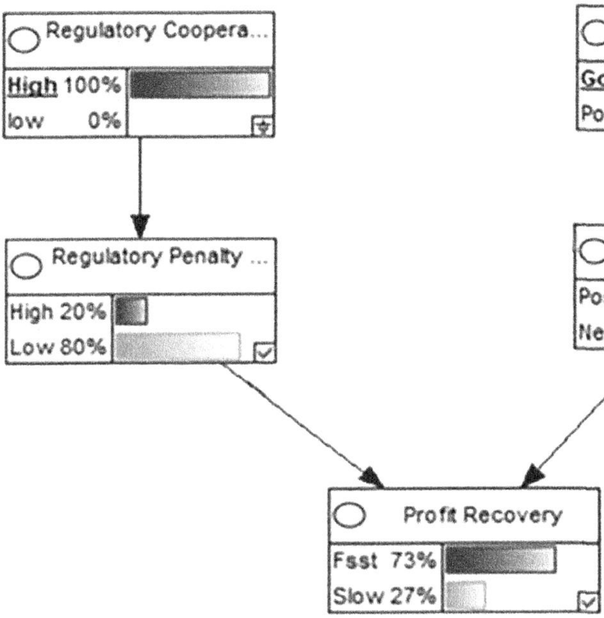

FIGURE 8.5 Bayesian Model – Best-Case Scenario.

explore the conditional relationships between actions and outcomes, assess alternative strategies, and quantify the consequences of each path.

Furthermore, the model's ability to update beliefs in response to new evidence supports adaptive decision-making critical in evolving crisis contexts. It also aligns with the principles of causal reasoning in complex systems, enhancing transparency and policy relevance. By visualizing these dependencies and outcome probabilities, the Bayesian approach supports both ethical reasoning and strategic alignment, making it a suitable choice for this high-stakes scenario.

8.2.2 Markov Decision Process

Chapter 7 introduces the Markov chain model and its applications in the decision-making process. However, this chapter explains in detail how the MDP can be applied in stochastic environments by addressing sequential problems. It emphasizes the requirements of making rational decisions, which can engage the careful consideration of next actions and activities. The MDP is applied as a discrete-time state transition model in making judgemental decisions in engineering applications.

MDPs employ mathematical equations that consider probabilistic calculations for decision optimization. These models are broadly utilized to design and develop intelligent systems that can deliver uninterrupted performance in uncertain situations.

In critical applications, the Markov network follows the MDP and its associated policy to deliver system operations. The MDP indicates the real-world scenario and the policy guides how the agent works properly in a Markov network. The Bellman equation follows Markov network theory, and it is considered as a fundamental equation to calculate the total Markov value:

$$\textbf{Value} = \textbf{V(S)} = \textbf{R(S)} + \Upsilon \, \Sigma \, \textbf{P(S' | S, A)} \, . \, \textbf{V(S')}$$

where the transition probabilities describe the dynamics of the world and represent P(S'|S, A): the probability of transitioning to state S' from state S under action A, R(S) represents the immediate value of rewards in state S, and Gamma Υ as loss or discount factor: $0 < \Upsilon < 1$.

In a Markov network, professionals need to identify states, actions, probability distribution, and rewards for instances and they can execute the Markov chain model with simulations. At each time sequence, the state can be monitored, and professionals can take actions based on the current state. They can select an initial state randomly from available options.

In different industrial applications, professionals can apply Markov models such as in traffic control systems, patient booking and admission systems, service optimization in telecom systems, stock exchange trading systems, and smart queue control systems in the manufacturing industry.

8.2.2.1 Use Case Scenario

This chapter applies the Markov model considering the Theranos scandal using cognitive, model-driven, and data-driven decision-making frameworks to identify root causes and propose preventive solutions. It reveals how flawed executive judgement, driven by psychological biases and poor oversight, led to ethical failure.

Theranos, founded in 2003 by Elizabeth Holmes, promised groundbreaking advances in medical diagnostics. It claimed its technology could accurately conduct hundreds of tests using just a few drops of blood from a finger prick. The startup quickly gained fame and financial backing, reaching a staggering valuation of $9 billion at its peak. However, Theranos' technology was fundamentally defective, leading to widespread misinformation, compromised patient safety, and massive financial losses once exposed.

The case unfolded primarily in Silicon Valley, California, within the heart of technological innovation and venture capitalism. Holmes leveraged Silicon Valley's culture of disruptive innovation and ambition to rapidly grow Theranos, attracting major investors like Rupert Murdoch and Betsy DeVos and forming partnerships with Walgreens and Safeway. Despite warnings and internal concerns about technology accuracy and reliability, Theranos aggressively pursued commercialization without proper scientific validation or transparency.

Theranos' fraudulent practices were exposed, resulting in severe consequences, including massive financial losses for investors, significant health risks for patients relying on faulty tests, and serious legal repercussions, notably Elizabeth Holmes' 11-year prison sentence. Furthermore, the scandal caused lasting reputational harm to Silicon Valley's biotech sector, prompting stricter regulatory scrutiny and heightened scepticism towards ambitious technological claims.

The Theranos case highlights three major decision-making failures: first, the deliberate misrepresentation of the technology's capabilities driven by overconfidence and optimism biases; second, significant ethical lapses in corporate governance, characterized by the suppression of whistleblowers and a culture of secrecy and groupthink; and finally, the manipulation of financial data to deceive investors, reinforced by cognitive biases such as anchoring, illusory truth effect, and strategic deception. These decisions collectively led to Theranos' dramatic collapse.

The effectiveness of a model-driven decision-making approach lies in its ability to provide a structured, evidence-based framework for analysing complex situations. In the case of Theranos, a key model – the Markov Decision Model – offers valuable insights into the flawed decision pathways that led to corporate failure.

Complementing this psychological view, the Markov Decision Model offers a structural framework that maps Theranos's decision pathway as a sequence of state transitions. In this model, each critical moment – such as launching unvalidated technology, ignoring internal warnings, deceiving the public, and eventually facing legal consequences – can be represented as a "state" with associated decisions, transition probabilities, and rewards or penalties. This process highlights how early decisions influenced future limitations, showing a clear path of escalating commitment and shrinking alternatives. The model also reveals how the company moved from high-reward states (e.g., media attention, investor confidence) to low-reward or absorbing states (e.g., legal collapse), emphasizing the consequences of failing to intervene at pivotal junctures.

A Markov decision-making model provides a structured framework to evaluate how decisions at each stage influenced Theranos's overall trajectory. In this model, each "state" represents a decision environment in which the company finds itself, and "transitions" are driven by choices made by leadership. Each action carries a reward or penalty, and probabilities indicate the likelihood of specific outcomes based on the decision made.

There are a few states, which are strongly associated with the use case scenario, and we can describe those states as follows:

S1	Launching a product with incomplete validation
S2	Receiving internal warnings about data inaccuracies
S3	Choosing to suppress whistleblowers
S4	Misleading stakeholders and continuing PR campaigns
S5	Legal investigation and exposure of fraud
S6	Court proceedings and company collapse

Considering the above-identified states in the Theranos Decision-Making Trajectory, an MDP map can be drawn as shown in Figure 8.6. It illustrates the sequential process from the launch state towards the system collapse.

A Quantitative research method is executed to identify the transition probability for each state and also to assess the respective reward points for each state. For better understanding of readers, we can assume arbitrary values (transition probability matrix as shown in Table 8.4 and reward value matrix as shown in Table 8.5) to

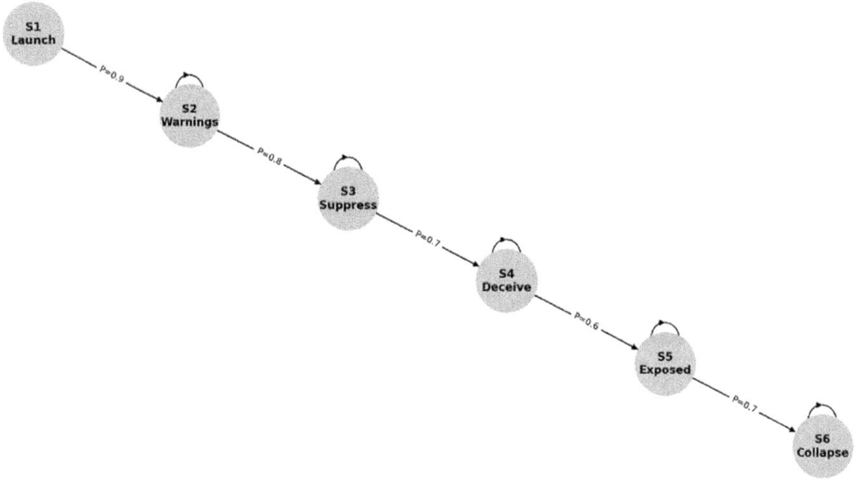

FIGURE 8.6 MDP Model for Theranos Decision-Making Trajectory.

TABLE 8.4
Transition Probability Matrix for States

From/To	S1	S2	S3	S4	S5	S6
S1	0	0.9	0.1	0	0	0
S2	0	0.1	0.8	0.1	0	0
S3	0	0	0.1	0.7	0.2	0
S4	0	0	0	0.2	0.6	0.2
S5	0	0	0	0	0.3	0.7
S6	0	0	0	0	0	1

validate the developed Markov model for the Theranos Decision-Making Trajectory incident.

Early-stage S1 offers short-term rewards, driven by investor support and media coverage, despite associated risks. Transitioning to S2 reflects emerging internal risks, which are often ignored (high P to S3). Continued suppression of dissent (S3) escalates deception (S4), increasing long-term penalties. Eventually, systemic failure leads to exposure (S5) and collapse (S6), where recovery is no longer feasible (absorbing state).

In an MDP, the V matrix (or value function V(S) represents the expected long-term reward for each state, assuming optimal decision-making from that state onward. The Bellman equation can be used to compute the process recursively. We can apply this equation to calculate the Markov value for the Theranos incident, considering the identified states. Let's assume that the discount factor for this situation is $\gamma=0.7$.

If the transition probabilities are defined as per your previous Markov chain (e.g., 90% chance from S1 to S2, 10% to S3), we substitute those into the Bellman function.

TABLE 8.5
Reward Value Matrix for States

State	Reward
S1	+5 (initial success/media hype)
S2	-2 (warnings; internal tension)
S3	-4 (morale decline; legal risk)
S4	-5 (deeper deception)
S5	-10 (fraud exposed, public scandal)
S6	-15 (legal collapse; imprisonment)

Updating each state value repeatedly until it converges (i.e., the change between iterations becomes very small):

$$V(s1) = R(s1) + \gamma - [P(s2|s1) - V(s2) + P(s3|s1) - V(s3)]$$

$$V(s2) = R(s2) + \gamma - [P(s3|s2) - V(s3) + P(s2|s2) - V(s2) + P(s4|s2) - V(s4)]$$

The Markov value for all states can be calculated following the above equation and the results can be obtained as shown in Table 8.6.

The results indicate a clear pattern of escalating consequences across decision stages. First, the sharp decline in expected values at each subsequent stage demonstrates the compounding effect of unethical or poorly governed decisions, where early missteps amplify negative outcomes over time rather than remaining isolated. Second, the model highlights the importance of early intervention: corrective actions at intermediate stages such as S2 or S3 – through mechanisms like acknowledging internal warnings or adopting transparent practices – could have significantly mitigated long-term damage. Finally, the apparent attractiveness of S1 is shown to be misleading. While the initial launch produces a short-term positive payoff, the results reveal that this gain is unsustainable and rapidly deteriorates into failure when validation and oversight are neglected.

This Markov approach captures the sequential deterioration of decision quality in Theranos and reveals how failure to act on early risk signals can entrench an organization in unethical behaviours. The probabilistic structure reflects real-world uncertainty, while the reward matrix shows the growing cost of unethical decisions. This model demonstrates how structured frameworks can simulate outcomes, highlight decision inflection points, and support preventive strategies.

8.3 RATIONALE OF MODEL APPLICATIONS – BAYESIAN AND MARKOV

A complex, high-risk engineering environment requires navigation through uncertainty, dynamic system behaviour, and significant human and organizational error potential. These environments are complex adaptive systems characterized by

TABLE 8.6
Markov's Result on Theranos Case Study

State	Meaning	Expected Value (V)	Interpretation
S1 (Launch)	Launch the product without full validation	**–11.25**	Although there is initial media hype and funding success, the long-term costs of launching prematurely outweigh the short-term gains.
S2 (Warnings)	Receive internal warnings about failures	**–22.53**	Ignoring early warning signs can lead to deeper entrenchment in risky behaviour and declining organizational health.
S3 (Suppress)	Suppress whistleblowers and dissent	**–29.45**	Covering up problems amplifies risk, destroys internal trust, and increases the likelihood of exposure.
S4 (Deceive)	Publicly promote false narratives	**–35.26**	The cost of continued deception grows – media and regulators begin to scrutinize, eroding credibility.
S5 (Exposed)	External parties uncover fraud	**–43.65**	At this point, irreversible damage has occurred. Legal, reputational, and financial consequences dominate.
S6 (Collapse)	Total organizational collapse	**–60.00 + (absorbing)**	The worst-case state, where company dissolution and criminal charges are inevitable. Future rewards no longer exist.

nonlinear interactions and feedback loops, which make prediction and control difficult. Moreover, cognitive limitations, biases, and system-level interdependencies can severely influence safety outcomes. A probabilistic framework for decision-making under uncertainty can be applied in rational decision-making, using computational models like BN and MDPs to identify optimal decisions. The aim of this chapter is to apply model-driven decision-making theories and models to analyse critical situations and to demonstrate how structured decision frameworks can enhance reliability and safety in high-risk engineering systems.

A model-driven method, especially BN, is highly suitable for a strategic and programmatic decision-making environment. BNs are probabilistic graphical models that excel at representing complex systems under uncertainty. Therefore, a BN is found to be the most appropriate data-driven model to analyse this decision, as it provides a formal model-driven framework that integrates this evidence and quantifies how each one influences the probability of the structural dependencies.

BNs are ideally suited for modelling situations characterized by uncertainty, interdependent risk factors, and incomplete information. In BNs, decision-makers can present and analyse probabilistic relationships between variables such as contractor safety history, hazard severity, and emergency equipment availability. Moreover, it supports belief updating, allowing the model to adjust its predictions based on new evidence (e.g., audit results or incident reports).

A BN provides a structured framework for rationally integrating the multiple, uncertain data streams. It demonstrates how quantitative and qualitative signals interact to alter the probability of success for each action. Using the GeNIe Modeler software, an influence diagram can be constructed to represent the statistical logic of the decision-making process. BN has directly and effectively can address critical situations through the following methods:

- Quantifying uncertainty: It combines a large amount of relevant knowledge and historical data to simulate the probability of corrosion failure and other events, in order to make corresponding predictions.
- Mapping interdependencies: It visually illustrates how factors such as inspection frequency, coating conditions, and soil corrosiveness collectively affect the risk of leakage through numbers and icons.
- Provide a quantitative basis for decision-making: It shifts management measures from intuitive cost reduction to risk assessment using historical data and model building. This method directly and effectively eliminates the drawbacks of cost-first heuristics.
- Enabling predictive analysis: BNs analyse different hypothetical scenarios to determine the impact of different investment strategies on overall system risk.

BNs provide a powerful means of supporting rational decision-making in environments characterized by elevated levels of uncertainty. This chapter models and analyses an emergency scenario involving complex conditions, using a representative example to illustrate the approach. Within this context, the analysis identifies plausible alternatives that can reduce uncertainty, which otherwise may contribute to incidents, delayed decision-making, poor coordination of actions, and an increased scale of casualties. By constructing a BN, professionals can infer posterior probabilities of casualty levels under varying conditions and assess how specific decisions influence expected outcomes, thereby quantifying both the effectiveness and risks of decisions made under uncertainty.

Markov chain model represents another category of stochastic sequential decision-making frameworks, where cost and transition functions are determined solely by the present state of the system and the action taken. This characteristic simplifies the complexity of decision-making by focusing on immediate state-action dependencies. These models find extensive application across diverse fields, particularly in queueing theory and inventory management, where they optimize resource allocation and operational strategies based on real-time system dynamics.

MDPs demonstrate the importance of a formal decision framework for optimizing choices under uncertainty through the systematic evaluation of actions and outcomes. When integrated with other perspectives, this approach offers a comprehensive understanding of critical situations by revealing how cognitive biases at the individual level and the absence of model-driven structures at the organizational level can jointly contribute to failure. The capacity of MDPs to represent the systemic reinforcement of risky behaviours and the misalignment between institutional incentives and long-term safety objectives provides a solid foundation for analysing and assessing complex, uncertainty-driven situations.

The MDP builds upon Markov chains by incorporating actions and rewards, enabling the evaluation of optimal decision policies under conditions of uncertainty. Within this framework, a crisis can be represented as a discrete-time state transition system, where states reflect economic or operational conditions, actions correspond to policy choices, and rewards capture the outcomes of those decisions. As a complementary approach, MDPs are particularly effective for analysing decision-making in complex adaptive systems. This model-driven framework helps professionals identify suboptimal choices that may persist over time and examines organizational defensive systems across multiple levels, from technical review to operational management. The model integrates inputs such as organizational structure, communication patterns, and historical incident data, while key assumptions include defensive independence and detection capability that can support the systematic management of uncertainty within organizational contexts.

Collectively, the model-driven application of BNs and MDPs offers a robust quantitative framework for examining how interacting variables shape uncertainty within complex systems. BNs employ probabilistic inference and sensitivity analysis to identify the most influential factors contributing to unsafe outcomes, while MDPs model potential state transitions from safe to critical conditions and support the timing of preventive actions, such as suspending operations or initiating crowd evacuation. Together, these decision-making models establish a comprehensive approach to public safety and risk management. When combined with cognitive insights, probabilistic modelling enables professionals to anticipate emerging hazards, develop effective early-warning mechanisms, and optimize real-time decision responses.

8.4 CONCLUSION

This chapter has illustrated the application of model-driven analysis using BNs and MDPs, demonstrating quantitatively how system resilience can be assessed and strengthened. By mapping the causal relationships between factors such as rainfall, communication system failures, decision-making speed, and resulting casualties, these models provide a rigorous framework for evaluating structural risks based on observable evidence. The analysis highlights that effective decision-making in complex and high-risk environments requires not only a clear understanding of these interdependencies but also a disciplined, data-informed approach to anticipate potential failures and mitigate their consequences.

Collectively, Bayesian and Markov models offer valuable insights into why decisions fail and how they can be improved by integrating behavioural intuition with structured, quantitative planning. When embedded into engineering and operational workflows, these frameworks enhance safety culture, sharpen risk anticipation, and guide professionals towards better-informed, timely actions. Ultimately, adopting these integrated decision-making models fosters a proactive, resilient organizational mindset, where cognitive awareness and predictive analytics work together to improve decision quality, enforce compliance, and prevent future incidents, particularly in high-stakes contexts such as confined space operations or high-rise developments.

BIBLIOGRAPHY

Bansal, T., King, M., & Seijts, G. (September 26, 2015). The Volkswagen emissions scandal: A case study in corporate misbehaviour. *The Globe and Mail*. www.theglobeandmail.com/report-on-business/rob-commentary/the-vw-emissions-scandala-case-study-in-what-not-to-do/article26550100/.

Bennett, C. C., & Hauser, K. (2013). Artificial intelligence framework for simulating clinical decision-making: A Markov decision process approach. *Artificial Intelligence in Medicine*, 57(1), 9–19. https://doi.org/10.1016/j.artmed.2012.12.003.

Boucherie, R. J., & van Dijk, N. M. (Eds.). (2017). *Markov decision processes in practice* (1st ed.). Springer International Publishing. https://doi.org/10.1007/978-3-319-47766-4.

Cambaza, E. (2024). Theranos: The ethics of innovation and deception in healthcare technology. *REVES – Revista Relações Sociais*, 7, 19792.

Fenton, N., & Neil, M. (2012). *Risk assessment and decision analysis with Bayesian networks*. CRC Press LLC. https://doi.org/10.1201/b21982.

Gal, Y., & Pfeffer, A. (2008). Networks of influence diagrams: A formalism for representing agents' beliefs and decision-making processes. *The Journal of Artificial Intelligence Research*, 33, 109–147. https://doi.org/10.1613/jair.2503.

Gates, G., Ewing, J., Russell, K., & Watkins, D. (March 16, 2017). How Volkswagen's 'defeat devices' worked. *The New York Times*. www.nytimes.com/interactive/2015/business/international/vw-diesel-emissions-scandal-explained.html.

Hotten, R. (December 10, 2015). Volkswagen: the scandal explained. *BBC News*. www.bbc.com/news/business-34324772.

Huang, L., Chen, T., Deng, Q., & Zhou, Y. (December 2023). Reasoning disaster chains with Bayesian network estimated under expert prior knowledge. *International Journal of Disaster Risk Science*, 14(6), 1011–1028. https://doi.org/10.1007/s13753-023-00530-w.

Kahneman, D., & Tversky, A. (1974). *Judgment under uncertainty: Heuristics and biases* (pp.1124–1131). Cambridge University Press.

Lee, P., Xiaogang, C., & Hajime, K. (2020). Willingness to pay for brand reputation: Lessons from the Volkswagen diesel emissions scandal. Working Papers. https://ideas.repec.org/p/cty/dpaper/20-02.html.

Moreira, C., Nguyen, N. L. C., Francis, G., Mohasel Afshar, H., Lopatnikova, A., Cripps, S., & Marchant, R. (2025). Bayesian causal discovery for policy decision making. *Data & Policy*, 7, Article e26. https://doi.org/10.1017/dap.2024.93.

Thomas, L. C. (1995). Markov decision processes: Markov decision processes: Discrete stochastic dynamic programming. *Journal of the Operational Research Society*, 46(6), 792–793. https://doi.org/10.1057/jors.1995.110.

9 Theory and Model Applications

Multi-Criteria Decision-Making

9.1 BACKGROUND

The foundations of modern decision-making theory can be traced to the seminal work of von Neumann and Morgenstern in 1944, who proposed that rational individuals select among alternatives by choosing the option that maximizes expected benefit according to their preferences (Neumann & Morgenstern, 1944). As decision-making contexts became increasingly complex, Keeney and Raiffa expanded the classical perspective by introducing the concept of multi-criteria decision-making (MCDM) in 1976 (Keeney & Raiffa, 1976). Their framework acknowledged that professional decisions often involve competing objectives and multiple evaluation criteria, providing a structured approach for achieving rational outcomes in environments where singular measures of value are insufficient. This work formalized MCDM by providing a structured framework for evaluating and balancing multiple, often conflicting objectives in complex decision contexts.

The MCDM constitutes the main ideas to enhance professionals' capabilities to figure out essential and critical factors, assign a weight to each factor, and turn those into clear, actionable insights by executing mathematical calculations. For example, when you would like to buy a smartphone, you can identify relevant factors such as weight, cost, battery longevity, durability, brand reliability, maintenance and update features, and customer care, which are sometimes critical to make decisions. Individual buyers can apply MCDM to select the best option from the available smartphones on the market.

MCDM is a very efficient decision-making tool in making decisions, where multiple criteria are available and their weightages are variables in nature (Dash et al., 2017). Professionals can apply MCDMs in their work environments to select the best option out of available alternatives. Sometimes, these are critical, but they can provide a higher level of acceptance and credibility from the senior leaders. There is a crucial process to select a suitable vendor to implement information systems, and it demands to follow a credential approach with proper justification. MCDM can allow professionals to follow sequential steps before making final decisions.

The MCDM analysis seeks to generate comparative ratings by aligning objectives with assigned weights and to identify a set of non-inferior solutions through trade-offs among competing objectives and benefits under uncertainty. A non-inferior

 DOI: 10.1201/9781003685449-9

solution represents an effective means of addressing complex problems by simultaneously considering multiple criteria and goals, enabling decision-makers to avoid suboptimal or high-risk choices. As a practical and robust approach, MCDM supports the identification of the most suitable solution in uncertain contexts. Decision-making within this framework typically follows a structured process that involves clarifying objectives, identifying feasible alternatives, defining evaluation criteria, systematically analysing and comparing options, selecting the preferred alternative, and validating the outcome through stakeholder review.

For instance, consider a selection committee tasked with choosing the most suitable vendor to implement a cybersecurity solution for a banking organization. The committee would first establish relevant evaluation criteria aligned with organizational goals and assign relative weights to reflect their importance. When issuing a public request for proposals, these criteria would be clearly articulated to guide both technical and financial submissions. Typical criteria might include solution cost, quality of service delivery, prior experience in deploying similar systems, and compliance with safety or security certifications. Each criterion can then be weighted according to organizational priorities – for example, 40% for cost, 25% for prior experience, 20% for safety certifications, and 15% for quality of service. After assessing the submitted proposals, the committee assigns scores to each vendor based on their performance against the criteria. If, for example, provider X receives a score of 8 out of 10, provider Y a score of 7, and provider Z a score of 7.5, the application of the MCDM framework would lead.

Professionals can follow MCDM models in making rational decisions for real-world scenarios that often have several objectives. The procedural steps of MCDM can be illustrated as shown in Figure 9.1. For example, autonomous driving needs to balance efficacy and obstacles in traffic movement as well as to address sociocultural effects. The objectives of autonomous driving can conflict with traditional traffic systems, and professionals can consider multidimensional factors to improve transportation systems by accommodating intelligent traffic systems.

The decision-making process begins by clearly defining the objective, such as selecting the most suitable supplier. Next, relevant criteria are established to determine what factors matter most in achieving this goal, for example, cost, quality, and delivery time. Alternatives are then generated, such as different suppliers, and each option is evaluated and ranked against the defined criteria. Finally, the alternative with the highest overall value is selected as the preferred decision. It allows professionals

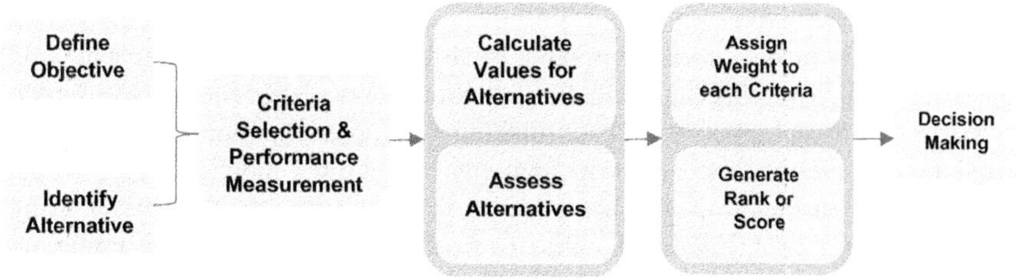

FIGURE 9.1 MCDM Process.

to follow data-driven models in making rational, fair, and judgemental decisions, where multiple criteria and contextual information are considerable factors. The MCDM model is considered to be the most suitable decision-making tool for real-world engineering practices, particularly in uncertain situations.

Professionals follow several types of Multi-Criteria Decision-Analysis (MCDA) techniques that can allow organizations to get rankings for alternatives and can choose the most preferred one. These analytical techniques are robust to manage complex situations that have multiple criteria and objectives. It presents several ways of measuring performance for each alternative, weighting the objectives, and calculating ranks for available options. Then, the analytical models intend to make rational decisions by choosing the most suitable option. In complex situations, decision-making would be more critical for professionals who engage in alternatives based on multiple criteria. The model delivers a systematic approach to making decisions by addressing alternatives, objectives, and criteria.

The MCDM undertakes that critical decision-making situations involve multiple objectives and criteria, such as costs, performance, quality, schedule, efficiency, workforce, environmental impact, features, and social effects. The fundamental principle is to balance these criteria for an optimal decision. The process involves defining the problem, identifying decision criteria, generating alternatives, evaluating these alternatives, and selecting based on the evaluations. This structured approach assists in assessing trade-offs among criteria, objectives, delivering rational decision-making.

9.2 MCDM METHODS

There are different MCDM methods that provide distinct ways to assess alternatives based on multiple criteria. This chapter explains two major established methods, which are popular in delivering rational decision-making.

9.2.1 Analytic Hierarchy Process

The Analytic Hierarchy Process (AHP) is a widely used and popular MCDA method that simplifies complex decision-making through structured pairwise comparisons. Rather than relying solely on numerical calculations, AHP enables professionals to compare pairs of criteria or alternatives to express their relative preferences. It breaks down complex problems into smaller, manageable components and evaluates criteria and alternatives using a systematic comparison technique to derive relative weights. Based on these weights and the resulting scores, decision-makers can identify the most suitable alternative.

AHP visually models decision-making by structuring the problem into a hierarchy consisting of the overall goal, evaluation criteria, and available alternatives (Guidi et a., 2020). This model supports decision-making in complex situations by decomposing problems into hierarchical subgroups that can be analysed independently. Through pairwise comparisons, it establishes a hierarchy of criteria and alternatives using a rigorous mathematical framework, allowing professionals to justify their preferences logically. An optimized hierarchy reduces the number of required comparisons while maintaining analytical consistency. By integrating criteria weights with alternative performance scores, AHP ranks the available options and determines final priorities. The method is particularly effective for problems involving multiple criteria and subjective judgements.

The AHP process begins by structuring the decision problem into a clear hierarchy that visually represents the relationships between the overall goal, the evaluation criteria, and the available alternatives. The criteria are then compared pairwise to capture the decision maker's relative preferences for assessing the alternatives. Next, the alternatives are compared two at a time under each criterion based on their relative performance. From these comparisons, criterion weight vectors are calculated using matrix transformations to produce normalized weights that reflect the importance of each criterion. Similarly, the pairwise comparisons of alternatives generate a matrix of normalized performance scores for each option. Finally, the criteria weights are combined with the alternative scores to rank the alternatives and identify the most preferred option.

9.2.1.1 Use Case Scenario – Boeing 737 Max Crisis

The Boeing 737 Max crisis represents one of the most consequential failures in modern aviation, defined by two fatal accidents: Lion Air Flight 610 on 29 October 2018 and Ethiopian Airlines Flight 302 on 10 March 2019 (NTSB, 2020). These tragedies exposed critical weaknesses in aircraft design, certification, and oversight, leading to the loss of 346 lives and raising serious concerns about safety governance in the global aviation industry.

Investigations revealed that both crashes were linked to the Manoeuvring Characteristics Augmentation System (MCAS), a software system introduced to compensate for aerodynamic changes caused by the larger LEAP-1B engines positioned further forward on the 737 Max. MCAS was intended to automatically lower the aircraft's nose when a high angle of attack was detected. However, the system relied on data from a single angle-of-attack sensor at a time, and in both accidents, faulty sensor readings repeatedly activated MCAS, overpowering pilots' attempts to maintain control. Boeing's strategic objective of minimizing additional pilot training to preserve commonality with earlier 737 models significantly influenced these design choices.

Following the first crash, the U.S. Federal Aviation Administration (FAA) permitted the 737 Max to remain in service, issuing only an Airworthiness Directive that required updates to flight manuals. It was not until after the second crash that aviation authorities worldwide took decisive action, grounding the entire 737 Max fleet on 13 March 2019. The scope focuses on decisions made between 2011 and 2019, drawing on available data, established engineering best practices, and recognized decision-making frameworks.

This chapter examines key decision points through a model-driven decision-making framework using the AHP method to evaluate Boeing's strategic responses to aerodynamic instability: (A) redesigning the airframe, (B) mandating extensive pilot training, or (C) deploying a software-based solution through MCAS.

The AHP, developed by Thomas Saaty, offers a systematic approach to complex decision-making by breaking a problem into a hierarchy of goals, criteria, and alternatives (Saaty, 1980). It uses pairwise comparisons to calculate relative weights and scores, while consistency in judgements is validated through the Consistency Ratio (CR). This structured methodology ensures that both quantitative and qualitative factors are evaluated in a transparent and logical manner.

In this case, the overall goal is to identify the safest and most effective solution to address the aerodynamic challenges of the Boeing 737 Max. The evaluation is based on four key criteria: safety impact, cost implications, regulatory compliance, and time-to-market, as shown in Table 9.1. Three strategic alternatives are assessed against these criteria: (A) redesigning the airframe, (B) implementing comprehensive pilot training, and (C) deploying a software-based solution through the MCAS system. It uses pairwise comparisons to establish a hierarchy of criteria and alternatives, allowing decision-makers to articulate their preferences in a structured manner.

AHP uses numerical scales to evaluate and compare options based on criteria, assigning numerical weights to each criterion for relative valuation. Mathematical algorithms combine the scores and weights for each option using a simple weighted average. The columns for normalization are shown in Table 9.2.

AHP uses various matrix transformations to determine criteria weight vectors from pairwise comparison results, reflecting normalized criteria weightings as shown in Table 9.3.

Then, pairwise comparisons within each criterion create a matrix of normalized performance scores for each alternative. Using the vectors of criteria weights and the matrix of alternative scores to determine each alternative's ranking as shown in Table 9.4.

The weighting of criteria reflects their relative importance in aviation decision-making. Safety impact is assigned the highest weight (0.54), recognizing that

TABLE 9.1
Pairwise Comparison Matrix (Criteria)

Criteria	Safety Impact	Cost Implications	Time-to-Market	Regulatory Compliance
Safety Impact	1	3	5	4
Cost Implications	1/3	1	3	2
Time-to-Market	1/5	1/3	1	1/2
Regulatory Compliance	1/4	1/2	2	1

TABLE 9.2
Total Columns for Normalization

Criteria	Total
Safety Impact	1.78
Cost Implications	4.83
Time-to-Market	11.00
Regulatory Compliance	7.50

TABLE 9.3
Normalized Matrix for Criteria

Criteria	Safety Impact (1.78)	Cost Implications (4.83)	Time-to-Market (11.00)	Regulatory Compliance (7.50)
Safety Impact	0.562	0.621	0.455	0.533
Cost Implications	0.185	0.207	0.273	0.267
Time-to-Market	0.112	0.068	0.091	0.067
Regulatory Compliance	0.141	0.103	0.182	0.133

TABLE 9.4
Priority Vector (Final Criteria Weights derived from Normalized Matrix)

Criteria	Priority Weight (Row Average)
Safety Impact	0.54
Cost Implications	0.23
Time-to-Market	0.09
Regulatory Compliance	0.14

catastrophic failure is unacceptable in commercial aviation. Cost implications (0.23) are important but remain secondary to safety considerations, as financial savings cannot justify increased risk. Regulatory compliance (0.14) supports sustained operational approval and mitigates legal exposure, while time-to-market (0.09), though relevant for competitiveness, is considered less critical than safety and regulatory assurance. Therefore, safety and regulatory assurance would be the most efficient way to avoid the incident.

The AHP model is used to explain how intuitive, experience-driven judgements may have led Boeing to initially underestimate the risks associated with MCAS. The model is applied to show how a structured and quantitative approach could have enabled a more transparent evaluation of trade-offs among safety, cost, and time. The model provides a balanced comparison of naturalistic and analytical decision-making, highlighting the value of integrating both approaches in complex, high-risk industries.

9.2.1.2 Use Case Scenario – Oil Spill Incident

The Laura D'Amato oil spill of 3 August 1999 stands as a major maritime emergency that offers important lessons on decision-making under conditions of pressure, complexity, and uncertainty (Marine Incident Investigation Unit, 2000). This chapter analyses key decision points arising from the incident and applies a model-driven decision-making approach to illustrate how the AHP can improve both understanding and effectiveness in the management of maritime emergencies.

The incident took place at the Shell Gore Bay terminal in Sydney Harbour, Australia's busiest port and a critical maritime gateway. On the evening of 3 August 1999, during the discharge of Murban light crude oil from the Italian tanker Laura D'Amato, approximately 294,000 litres (about 300 tonnes) of crude oil were released into the harbour between 6:26 p.m. and 6:50 p.m. The spill occurred through open sea chest valves, despite established procedures designed to prevent such incidents. The crew mistakenly believed the valves were closed, relying on the presence of seals rather than physically verifying valve positions.

This failure led to the largest ship-based oil spill recorded in Sydney Harbour and prompted an unprecedented emergency response involving more than 500 personnel from over 30 public and private organizations. The incident resulted in significant environmental risk, substantial financial costs, regulatory changes, operational disruption, and long-term reputational impacts for maritime stakeholders. Nevertheless, swift action by Sydney Ports Corporation and Shell personnel successfully confined the spill to Gore Cove and Balls Head Bay, enabling the recovery of an estimated 90% of the released oil and limiting the extent of environmental damage.

This use case focuses on a critical decision situation that examines the Pre-incident Valve Verification Protocol Selection, which proved critical to the incident's causation. This decision situation involved determining the appropriate verification methods for ensuring sea chest valves were properly closed before cargo discharge operations. The investigation revealed significant shortcomings in the verification protocols, as both ship and shore personnel failed to physically confirm the closed status of the relevant sea chest valve, instead relying on assumptions based on valve seals. Table 9.5 summarizes the key stakeholders and their functions in this decision-making situation. Developed by Saaty in 1971, the AHP is employed here as it offers a structured technique for analysing complex decisions involving an MCDM. AHP breaks down the decision problem (selecting verification protocols) into a hierarchy and uses pairwise comparisons to systematically evaluate how well each alternative protocol performs against criteria such as safety, efficiency, regulatory compliance, and technical feasibility, thereby quantifying subjective judgements and objective data to determine priorities.

The AHP provides an ideal framework for valve verification protocol selection through its structured approach to complex multi-criteria problems. AHP's pairwise comparison methodology helps to quantify qualitative judgements across four critical criteria: safety assurance, operational efficiency, regulatory compliance, and technical feasibility. This structured approach addresses the critical trade-offs in the Laura D'Amato incident (main event), where safety verification was compromised. The consequences included significant environmental damage, operational disruptions, regulatory investigations, substantial financial costs, and reputational harm to maritime stakeholders. Therefore, the purpose of this model is to showcase how hierarchical decomposition enables systematic evaluation of valve verification alternatives (hydrostatic, pneumatic, visual, ultrasonic testing) against these weighted criteria reflecting stakeholder priorities (see Table 9.5).

In marine applications, AHP has been used to select the best available technique for oil spill clean-up in marine protected areas and to assess the hazard levels to reevaluate decision issues. AHP has also been used to model future valve failures.

TABLE 9.5
Stakeholders Involved in Valve Verification Protocol

Stakeholder	Role/Action related to verification protocol	Key Interest related to Verification Protocol	Assessed Power	Assessed Interest
Ship Management/ Operators (e.g., owner – D'Amato Armatori)	Responsible for ship's Safety Management System (SMS) procedures under the International Safety Management (ISM) Code; Implemented/oversaw the protocol relying on checklist confirmation and ship crew checks.	Balancing operational efficiency and cost with regulatory compliance and vessel safety.	High (authority over defining and enforcing shipboard procedures via SMS)	High (responsible for vessel safety, liability
Vessel Crew (Master, Chief Mate, Pumpman)	Required by SMS/Pollution Prevention Manual (PPM) to perform physical/pressure tests but failed to do so adequately preloading/discharge; confirmed valve status on checklists, based on assumption/seals.	Having clear, understandable, and workable procedures that ensure personal safety without excessive workload.	Low (had limited influence on how the procedure was defined)	High (directly performed the task and were affected by safety outcomes)
Independent Cargo Surveyor(s)	Performed valve checks at loading port; interpreted requirements/valve status incorrectly (assumed closed); applied seals.	Accurately fulfilling duties per contract/standards; managing liability.	Medium (incorrect check at loading port perpetuated the hazard)	High (directly responsible for the verification task they were assigned)
Shell Terminal Staff (Shore Officer)	Completed International Safety Guide for Oil Tankers and Terminals (ISGOTT) Ship/Shore checklist by accepting ship is (Mate's) confirmation of valve status; followed terminal procedures.	Efficient vessel turnaround; terminal safety; verifying vessel compliance (via checklist).	Medium (accepted ship's verification status per procedure)	High (terminal safety, efficiency, incident impact)
Regulatory Bodies (Australian Transport Safety Bureau [ATSB], AMSA, Class Societies)	Set overarching standards (ISM Code, International Convention for the Prevention of Pollution from Ships (MARPOL), Class Rules, ISGOTT guidelines); investigated protocol failure post incident (e.g., ATSB, AMSA).	Ensuring adherence to established maritime safety standards and preventing incidents through effective regulation and oversight.	High (authority to set standards and investigate/penalize rule-breaking)	High (responsible for maritime safety oversight and investigation)

For valve verification, this transparency would highlight the dangerous imbalance between efficiency and safety verification, which is identified as a root cause. AHP's consistency-checking mechanism safeguards against logical inconsistencies in verification approaches contributing to the spill. AHP is chosen because it is simpler, directly calculates weights from judgements, and manages subjective perceptions well. The primary theoretical foundation of AHP lies in its hierarchical structure that consists of three distinct levels according to Saaty, which can be applied for the purposes of choosing a suitable valve verification protocol:

Level 1: Goal – At the top of the hierarchy sits the overall goal: "Select optimal valve verification protocol." This represents the ultimate decision objective – identifying the most appropriate method to verify valve status and functionality on maritime vessels to prevent incidents like the Laura D'Amato spill.

Level 2: Criteria – The second level contains the criteria against which verification protocols are evaluated:

- Safety Assurance: The protocol's ability to reliably detect valve failures and prevent hazardous situations
- Operational Efficiency: The protocol's impact on vessel operations, including time requirements and resource utilization
- Regulatory Compliance: The protocol's alignment with maritime regulations and industry standards
- Technical Feasibility: The protocol's practicality for implementation given vessel constraints and crew capabilities

Level 3: Alternatives – The bottom level consists of the verification protocol alternatives:

- Hydrostatic Testing: Applying liquid pressure to verify valve integrity and function
- Pneumatic Testing: Using air or gas pressure to verify valve sealing and operation
- Visual Inspection: Direct observation of valve position and condition
- Ultrasonic Testing: Using sound waves to detect internal valve conditions and potential leaks.

AHP employs pairwise comparisons on a 1–9 scale, as explained by Saaty, to convert qualitative judgements into quantitative priorities (Saaty, 1990). This allows maritime stakeholders to demonstrate the relative importance of criteria and alternatives, using Saaty's 1–9 scale to create pairwise comparisons – for example, in the Safety Assurance pairwise comparison matrix shown in Table 9.6.

- Hydrostatic Testing compared to Pneumatic Testing = 9 (extreme importance)
- Hydrostatic Testing compared to Visual Inspection = 3 (moderate importance)
- Hydrostatic Testing compared to Ultrasonic Testing = 5 (strong importance)

The same method was applied to the rest of the criteria in level 2: Operational Efficiency (Table 9.7), Regulatory Compliance (Table 9.8), Technical Feasibility (Table 9.9), and the Alternatives in level 3 (Table 9.10).

TABLE 9.6
Safety Assurance Pairwise Comparison Matrix

Safety Assurance	Hydrostatic Testing	Pneumatic Testing	Visual Inspection	Ultrasonic Testing	Avg
Hydrostatic Testing	1	9	3	5	0.57
Pneumatic Testing	1/9	1	1/7	1/3	0.05
Visual Inspection	1/3	7	1	3	0.27
Ultrasonic Testing	1/5	3	1/3	1	0.11
Sum =	1.644	20.000	4.476	9.333	

TABLE 9.7
Operational Efficiency Pairwise Comparison Matrix

Operational Efficiency	Hydrostatic Testing	Pneumatic Testing	Visual Inspection	Ultrasonic Testing	Avg
Hydrostatic Testing	1	1/3	2	1/3	**0.14**
Pneumatic Testing	3	1	5	2	**0.47**
Visual Inspection	1/2	1/5	1	1/3	**0.09**
Ultrasonic Testing	3	1/2	3	1	**0.30**
Sum =	**7.500**	**2.033**	**11.000**	**3.667**	

TABLE 9.8
Regulatory Compliance Pairwise Comparison Matrix

Regulatory Compliance	Hydrostatic Testing	Pneumatic Testing	Visual Inspection	Ultrasonic Testing	Avg
Hydrostatic Testing	1	1/2	1/3	2	**0.17**
Pneumatic Testing	2	1	1/2	3	**0.28**
Visual Inspection	3	2	1	3	**0.44**
Ultrasonic Testing	1/2	1/3	1/3	1	**0.11**
Sum =	**6.500**	**3.833**	**2.167**	**9.000**	

A key mathematical strength of AHP lies in its eigenvalue calculations – a method that determines the relative weights of criteria and alternatives. When maritime stakeholders make pairwise comparisons, these judgements form a matrix. The eigenvalue calculation processes this matrix to find the most stable set of weights that represent the stakeholders' priorities. For each matrix, the calculated priority vectors are obtained using the eigenvalue method. Using the Safety Assurance matrix as an example here:

- First, calculating column sums

TABLE 9.9
Technical Feasibility Pairwise Comparison Matrix

Technical Feasibility	Hydrostatic Testing	Pneumatic Testing	Visual Inspection	Ultrasonic Testing	Avg
Hydrostatic Testing	1	2	3	1/2	**0.27**
Pneumatic Testing	1/2	1	2	1/3	**0.16**
Visual Inspection	1/3	1/2	1	1/5	**0.09**
Ultrasonic Testing	2	3	5	1	**0.48**
Sum =	**3.833**	**6.500**	**11.000**	**2.033**	

TABLE 9.10
Alternatives Pairwise Comparison Matrix

Alternative Criteria	Safety Assurance	Operational Efficiency	Regulatory Compliance	Technical Feasibility	Avg
Safety Assurance	1	4	3	2	**0.47**
Operational Efficiency	1/4	1	1/2	1/2	**0.11**
Regulatory Compliance	1/3	2	1	1/2	**0.17**
Technical Feasibility	1/2	2	2	1	**0.25**
Sum =	**2.083**	**9.000**	**6.500**	**4.000**	

TABLE 9.11
Normalized Criteria for Safety Assurance matrix

Normalized Criteria Safety Assurance	Hydrostatic Testing	Pneumatic Testing	Visual Inspection	Ultrasonic Testing	Avg
Hydrostatic Testing	0.608	0.450	0.670	0.536	**0.57**
Pneumatic Testing	0.068	0.050	0.032	0.036	**0.05**
Visual Inspection	0.203	0.350	0.223	0.321	**0.27**
Ultrasonic Testing	0.122	0.150	0.074	0.107	**0.11**

Column sums: [1.64, 20.00, 4.48, 9.33], see last row of Table 9.6.

- Normalizing each column by dividing by its column sum

First column: [1/1.64, (1/9)/1.64, (1/3)/1.64, (1/5)/1.64], check Table 9.11 for decimal values for the entire matrix as Normalized Criteria for Safety Assurance matrix. However, in the Excel file, this was embedded in the formula for the "Average" column.

TABLE 9.12
Average Criteria (Priority Vectors for each Alternative across Criteria)

Average Criteria	Safety Assurance	Operational Efficiency	Regulatory Compliance	Technical Feasibility
Hydrostatic Testing	0.57	0.14	0.17	0.27
Pneumatic Testing	0.05	0.47	0.28	0.16
Visual Inspection	0.27	0.09	0.44	0.09
Ultrasonic Testing	0.11	0.30	0.11	0.48

Calculating row averages to obtain priority vectors:

Safety Assurance weights: [0.57, 0.05, 0.27, 0.11], see last column of Table 9.6
Calculating the Average (Priority Weight): (0.608 + 0.450 + 0.670 + 0.536)/ 4 = 2.264/4 = 0.566 ≈ 0.57

Criteria weights have also been obtained using all the steps above: [0.47, 0.11, 0.17, 0.25], see the last column of Table 9.10.

This approximates the solution to the eigenvalue equation $A_w = \lambda_{max}$ w, where w is the priority vector and λ_{max} is the principal eigenvalue.

After obtaining all the average criteria values, Table 9.12 is populated to show the priority vectors (average values) for each alternative across different criteria.

These values represent how well each testing method performs for each criterion – for example, Hydrostatic Testing scores 0.57 for Safety Assurance but only 0.14 for Operational Efficiency.

Criteria Weights then show the relative importance of each criterion in the decision-making process [0.47, 0.11, 0.17, 0.25].

Safety Assurance (see Table 9.10) has the highest weight (0.47), indicating it is the most important criterion, followed by Technical Feasibility (0.25).

Finally, the formula @MMULT was used for Matrix Multiplication of the final scores (see Table 9.13) by multiplying the alternative priority vectors by the criteria weights:

Final Score = Alternative Priority Vector × Criteria Weights

For Hydrostatic Testing:

$$= (0.57 \times 0.47) + (0.14 \times 0.11) + (0.17 \times 0.17) + (0.27 \times 0.25) = \mathbf{0.3789}$$

Alternative Scores and Rankings: The final scores determine the ranking of alternatives, where Hydrostatic Testing ranks No. 1 with a score of 0.3789, followed by Visual Inspection (No. 2), Ultrasonic Testing (No. 3), and Pneumatic Testing (No. 4).

TABLE 9.13
Criteria Ranks and Alternative Score

Criteria	Criteria Weights	MatrixMult.	Alternative Scores	Alternative Ranking
Safety Assurance	0.47	0.3789	**Hydrostatic Testing**	**#1**
Operational Efficiency	0.11	0.1600	**Pneumatic Testing**	**#4**
Regulatory Compliance	0.17	0.2347	**Visual Inspection**	**#2**
Technical Feasibility	0.25	0.2264	**Ultrasonic Testing**	**#3**

This hierarchical framework transforms valve verification from a fragmented process into a structured decision system where stakeholder priorities are explicitly weighted and verification alternatives are systematically evaluated against well-defined criteria.

The AHP analysis indicates that Hydrostatic Testing (0.3789) is the most effective verification method, with Visual Inspection (0.2347), Ultrasonic Testing (0.2264), and Pneumatic Testing (0.1600) following in order of effectiveness. These findings directly address the critical failures in the Laura D'Amato incident by prioritizing methods that provide definitive valve status verification rather than relying on assumptions.

9.2.2 Technique for Order Preference by Similarity to Ideal Solution

To compare the criteria and objectives considering the weightage, professionals can apply an MCDM model like the TOPSIS (Technique for Order Preference by Similarity to Ideal Solution) method, which is an efficient technique. This multi-criterion decision-making method evaluates how close each model is to the ideal solution, using weighted criteria relevant to disaster decision-making contexts (Behzadian et al., 2012). This method identifies alternatives based on their distance to an ideal solution and a negative ideal solution. The alternatives closest to the ideal solution and furthest from the negative ideal are considered the best choices.

TOPSIS integrates two sets of variables (criteria): Benefit and Cost criteria. The more is considered better in the set of benefit criteria (more is better), whereas less is considered better in the set of cost criteria. The TOPSIS hypothesizes two artificial solutions (alternatives) such as (i) positive Ideal solution: the one which has the best level for all attributes considered and (ii) negative ideal solution: the one which has the worst attribute values. Basically, the principle of TOPSIS is to select the alternative that is the closest to the ideal solution and farthest from the negative ideal alternative.

TOPSIS method is a broadly used in multi-criteria decision analysis that facilitates professionals to make accurate and informed decision-making in different industries. It encompasses to compare and assess multiple alternatives based on identified criteria. The TOPSIS picks the option closest to the ideal solution and farthest from the worst. This method aims to attain a balance between multiple criteria, narrowing down shortcomings in one aspect with strengths in another. Moreover, it grants a holistic modelling approach that doesn't limit potential solutions in critical and uncertain situations. Professionals can follow seven (07) steps in the TOPSIS, and these are described as a process flow map in Figure 9.2.

9.2.2.1 Use Case Scenario – Hydrogen Chloride Release at Wacker Polysilicon

The case study examined in this chapter focuses on the equipment failure and hydrogen chloride release that occurred on 13 November 2020 at Wacker Polysilicon North America. During maintenance activities at the company's facility in Charleston, Tennessee, a graphite heat exchanger fractured, leading to the release of hydrogen chloride (HCl) gas (U.S. Chemical Safety and Hazard Investigation Board, 2023). The incident took place on the fifth floor of an equipment access structure when a contractor pipefitter applied excessive torque while tightening flange bolts on the outlet pipe of the heat exchanger, which contained HCl. This over-tightening caused the component to crack, resulting in the unintended release of the hazardous gas.

At the time of the hydrogen chloride release, seven workers from two contracting companies were present on the platform, which had only a single staircase serving as both the access and exit route. Three of the workers, assigned to bolt-torqueing tasks, were wearing full-body chemical-resistant suits, while the remaining four, responsible for insulation work, were equipped only with standard flame-resistant clothing. The sudden release of HCl formed a dense white cloud that severely impaired visibility and blocked access to the staircase. In an attempt to escape, three workers climbed down external piping on the side of the structure, approximately 70 feet above ground level. All three fell during the descent, resulting in one fatality and severe injuries to the other two. The remaining four workers stayed on the platform until the release subsided and then evacuated safely using the staircase. The incident also caused approximately $214,000 in property damage at the Wacker facility.

This chapter applies the TOPSIS decision-making model to the situations encountered by Wacker that led to the HCl release and the damage that it caused. This allows professionals to develop solutions that could have prevented the incident and reduced its impact. By analysing the use case scenario through structured decision-making tools, the chapter aims to demonstrate how systematic evaluation can improve risk management in high-hazard industries.

The scenario highlights Wacker's inadequate emergency egress planning. Despite prior warnings, the company did not install a secondary exit on the fifth-floor platform where maintenance activities were conducted. When the hydrogen chloride release occurred, the sole staircase became unusable, leaving workers with no safe escape route and forcing some to descend via external piping, which resulted in one fatality and two serious injuries. At the time of the incident, the platform was served by only one egress staircase.

A Process Hazard Analysis conducted by Wacker in August 2020 had already identified the risk of personnel becoming trapped during a gaseous HCl release and recommended the installation of an additional exit, such as a ladder with an enclosed

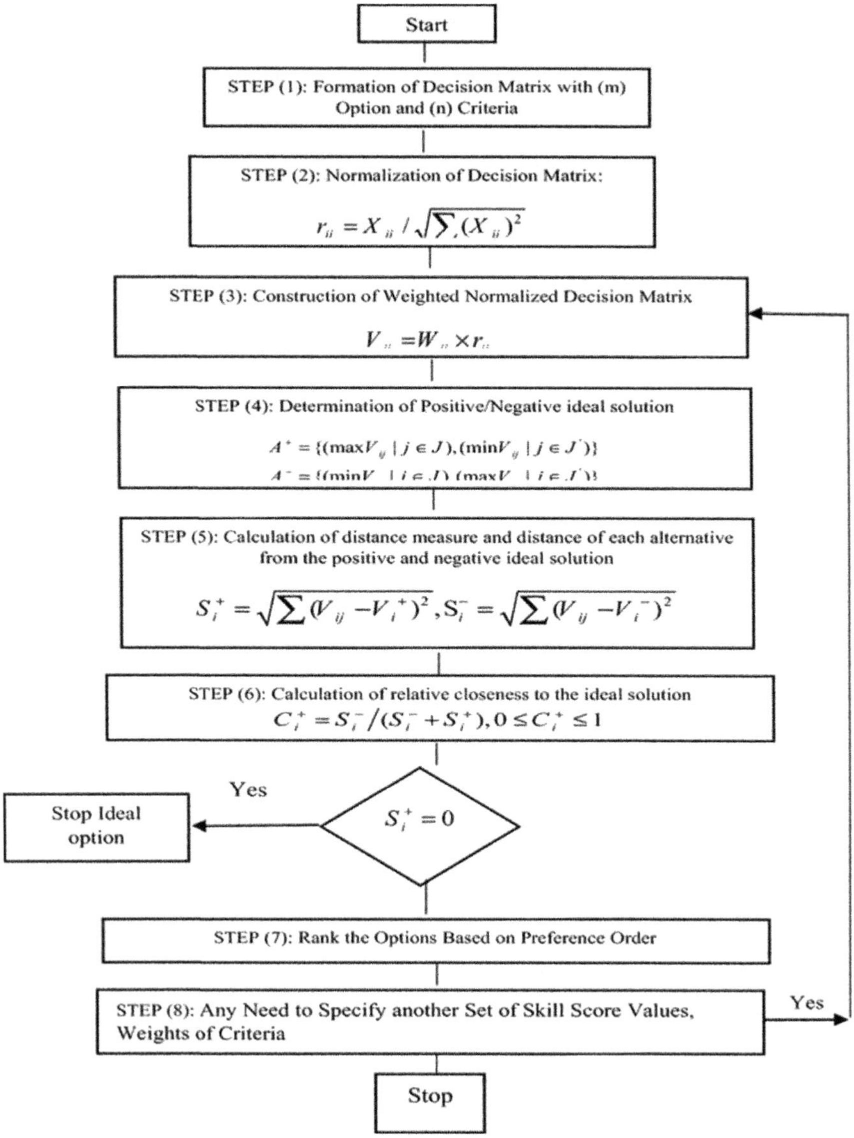

FIGURE 9.2 Process Flow Map on Seven Steps in TOPSIS.

safety cage. However, no alternative egress was implemented before the incident. Workers had also raised concerns regarding the lack of emergency exits, but these were dismissed by management, which maintained that existing arrangements complied with International Building Code requirements.

When the HCl release happened, the workers on the platform had only one escape option, but the staircase was then unreachable because of the HCl cloud's opacity.

Three of the four Pen Gulf workers began to climb down piping on the side of the equipment access structure, approximately 70 feet above the ground. All three fell, with one fatally injured and the others suffering serious injuries. The failure to provide adequate emergency egress was a significant oversight that increased the severity of the incident.

To analyse this decision-making situation, the chapter applies TOPSIS, which stands for Technique for Order of Preference by Similarity to Ideal Solution. The TOPSIS technique is a widely utilized multi-criteria decision analysis approach. The fundamental principle of TOPSIS is based on the fact that the best alternative is the one that is the closest to the ideal solution (the one that maximizes all the positive criteria) and the furthest from the anti-ideal solution (the one that has the weakest performances across all criteria).

In this case, the objective is to choose the best emergency egress improvement for the fifth-floor platform, considering multiple alternatives and various criteria. To do so, we considered various alternatives and evaluated them depending on different criteria, which were weights based on their importance.

The HCl incident at Wacker showed that having only one staircase on the platform severely limited evacuation options, which caused three workers to try to escape by climbing down piping on the side of the structure, leading to one fatality and severe injuries to the other workers. The following improvements could have been considered to reduce risks:

Alternative 1 (A1): Installing a second staircase
Alternative 2 (A2): Installing a caged escape ladder
Alternative 3 (A3): Installing an emergency slide chute
Alternative 4 (A4): Installing portable emergency ladders

With Alternative 1, the fifth-floor platform would have a second permanent egress route, ensuring a quick and secure evacuation even if the first staircase is obstructed. Alternative 2 offers a more compact evacuation route, adapted to more narrow spaces. The emergency slide chute of Alternative 3 allows for a quick descent and can be used even in case of low visibility. Finally, Alternative 4 can be considered because it can be advantageous in smaller spaces.

Step 1: Decision Matrix and Criteria Weighting

For the TOPSIS method, we considered four criteria. Table 9.14 provides the criteria and their weight.

The weights have been assigned based on the relative importance of each criterion in the context of the Wacker incident, as shown in Table 9.15. Safety is the most important aspect because the main objective is to prevent human losses and injuries, which is why it has been given the weight of 0.4. Cost, compliance, and feasibility have each been given the weight of 0.2. They are factors that influence the decision; however, no financial, regulatory or technical compromise can justify risking workers' safety.

TABLE 9.14
Definitions of Criteria with Weight

Criterion	Description	Weight
Safety Improvement (C1)	How much the solution improves evacuation safety	0.2
Cost (C2)	Installation and maintenance cost	0.2
Compliance (C3)	Alignment with updated building and safety codes	0.2
Feasibility (C4)	Practical ease of installation and maintenance	0.2

TABLE 9.15
Weighted Matrix for Alternative and Criteria

	C1	C2	C3	C4
A1	5	5	5	3
A2	3	4	4	5
A3	2	3	2	3
A4	2	4	3	4

Step 2: Normalized Decision matrix

To build the decision matrix, we gave a mark to each alternative regarding each criterion, between 5 (best option) and 1 (worst option).

To make the scores comparable across all criteria, vector normalization was applied. Each raw value was divided by the square root of the sum of squares in its column:

$$\text{Normalized Value} = \frac{Value}{\sqrt{\Sigma(\;column\ values)^2}}$$

This step assures that all criteria contribute equally to the analysis process, regardless of their original scales. We normalize the decision matrix using the normalized value formula as shown in Table 9.16.

Step 3: Weighted Normalized Decision matrix

The normalized values were multiplied by their respective weights to reflect the importance of each criterion. This step ensures that more critical factors, such as decision accuracy, have a greater impact on the final ranking.

$$\text{Weighted Value} = \text{Normalized Value}-\text{Weight}$$

We multiplied each column by the corresponding criterion's weight, then we determined the values of the ideal-worst and ideal-best options (Table 9.17).

TABLE 9.16
Normalized Matrix for Alternative and Criteria

	C1	C2	C3	C4
A1	0.772	0.615	0.680	0.391
A2	0.463	0.492	0.544	0.651
A3	0.309	0.369	0.272	0.391
A4	0.309	0.492	0.408	0.521

TABLE 9.17
Weighted Normalized Matrix

	C1	C2	C3	C4
A1	0.309	0.123	0.136	0.078
A2	0.185	0.098	0.109	0.130
A3	0.123	0.074	0.054	0.078
A4	0.123	0.098	0.082	0.104
Ideal-Worst	0.123	0.074	0.054	0.078
Ideal-Best	0.309	0.123	0.136	0.130

This matrix captures each model's weighted performance across the disaster-relevant criteria.

Step 4: Ideal and Negative-Ideal Solutions
TOPSIS compares each model to two benchmarks:

- Ideal Solution: Best performance across criteria (maximum for benefit criteria, minimum for cost criteria).
- Negative-Ideal Solution: Worst performance across criteria (minimum for benefit criteria, maximum for cost criteria).

These benchmarks define the performance limits for each criterion and are used in the next step to calculate distances.

Step 5: Distance to Ideal and Negative-Ideal Solutions
In this analysis, all criteria are treated considering both cost and benefit criteria (as shown in Table 9.18). E ternative is from both ideal and negative-ideal solutions. We calculated these distances using the following formula:

$$Distance = \sqrt{\Sigma}(weighted\ value - ideal\ worst\ or\ best)^2$$

TABLE 9.18
Ideal and Negative Solution Matrix

	C1	C2	C3	C4	Si-neg	Si-pos
A1	0.074	0.020	0.022	0.026	0.0293	0.0000
A2	0.049	0.015	0.011	0.016	0.0000	0.0293
A3	0.049	0.020	0.016	0.021	0.0090	0.0258
A4	0.049	0.015	0.011	0.016	0.0000	0.0293
Ideal-Worst	0.049	0.015	0.011	0.016		
Ideal-Best	0.074	0.020	0.022	0.026		

TABLE 9.19
Matrix for Relative Closeness and Rank

	Si-neg	Si-pos	Si-neg+Si-pos	Ci	Rank
A1	0.0293	0.0000	0.0293	1	**1**
A2	0.0000	0.0293	0.0293	0	**3**
A3	0.0090	0.0258	0.0348	0.2585187	**2**
A4	0.0000	0.0293	0.0293	0	**3**

These results show that the Bayesian Network model is much closer to the ideal solution and farther from the negative-ideal, indicating stronger overall performance based on the selected criteria.

Step 6: Closeness Score and Final Ranking
The last step calculates the closeness score, which reflects how near each model is to the ideal solution. It is computed as:

$$\text{Score} = \frac{\text{Distance to Negative}}{\text{Distance to Ideal} + \text{Distance to Negative}}$$

Then, we calculated the relative closeness of each alternative to the ideal solution. Finally, we were able to rank the alternatives, with 1 being the best solution as shown in Table 9.19.

The TOPSIS method allows us to draw some significant conclusions regarding the different possible evacuation options for the fifth-floor platform. This MCDM approach sheds light on the relative effectiveness of each alternative, considering hierarchized criteria, which are safety improvement, cost of installation and maintenance, compliance with building and safety codes, and practical feasibility.

Between the four alternatives that were considered, Alternative 1, which is installing a second staircase, appears to be the best one. A1 not only is the best option in terms of workers' safety and compliance with standards but also ensures permanent access to a second egress point, including in case of chemical smoke and/or reduced visibility. However, this option is the most expensive one and its feasibility is limited, especially because of space constraints on the platform.

Alternative 2 (caged escape ladder) is less expensive and easier to install than a staircase, and it aligns with codes. However, its safety is not as good in situations of high stress or for workers wearing bulky equipment. While Alternative 3 (emergency slide chute) allows for a quick descent that does not require any physical effort, it raises technical issues (space constraints and secure fastening) and its compliance with building and safety standards can be questioned. Finally, Alternative 4 (portable emergency ladders) is the least expensive option and also the easiest to implement. However, it has serious limitations in regard to safety, stability and compliance.

The TOPSIS model shows that Wacker's inaction does not result from a lack of practical solutions, but rather from a lack of a structured decision-making process. This analysis highlights the importance of taking educated decisions that are justified by multi-criteria data, especially in areas where human safety is at stake. Such an approach could have prevented or significantly mitigated the severity of Wacker's incident.

9.3 CONCLUSION

MCDM models have advantages over informal judgements as they are open and explicit, use established techniques for scores and weights, allow outsourcing of performance measurement, and provide a means of communication and an audit trail. The MCDM models use numerical scales to evaluate and compare options based on criteria, assigning numerical weights to each criterion for relative valuation. Mathematical algorithms combine the scores and weights for each option using a simple weighted average.

This chapter explains major decision failures from the identified use case scenarios using structured decision-making models. The AHP and TOPSIS were used to analyse the misinterpretation of the negative pressure test, providing a probabilistic view of how uncertain conditions led to risky decisions. These cognition-driven models (CDM) address the failure to maintain situations, focusing on criteria and their respective weightage.

AHP and TOPSIS methods were applied using weighted criteria such as accuracy, ease of use, and real-world applicability. Those models support to present ranks, which is highly effective for an organization in making rational decisions and also, these CDMs stand out for its simplicity and insight into behavioural risks. While no single model is universally superior, the chapter shows how AHP and TOPSIS can be applied in real-world applications, and the way of structured analysis in high-risk scenarios. Together, these models offer complementary strengths, supporting better decision-making, risk reduction, and resilience in complex, safety-critical environments.

BIBLIOGRAPHY

Behzadian, M., Khanmohammadi Otaghsara, S., Yazdani, M., & Ignatius, J. (2012). A state-of-the-art survey of TOPSIS applications. *Expert Systems with Applications*, *39*(17), 13051–13069. https://doi.org/10.1016/j.eswa.2012.05.056

Dash, S., Tripathy, B. K., & Rahman, A. ur. (2017). A holistic-based multi-criterion decision-making approach for solving engineering sciences problem under imprecise environment. In B. K. Tripathy, A. ur Rahman, & S. Dash (Eds.), *Handbook of research on modeling, analysis, and application of nature-inspired metaheuristic algorithms* (pp. 298–330). IGI Global. https://doi.org/10.4018/978-1-5225-2857-9.ch015

de Magalhães, R. F., Danilevicz, Â. de M. F., & Palazzo, J. (2019). Managing trade-offs in complex scenarios: A decision-making tool for sustainability projects. *Journal of Cleaner Production*, 212, 447–460. https://doi.org/10.1016/j.jclepro.2018.12.023.

Federal Aviation Administration (FAA). (November 7, 2018). Continued airworthiness notification to the international community: Boeing model 737-8 and 737-9 (737 MAX). www.faa.gov/sites/faa.gov/files/2021-08/Continued%20Airworthiness%20Notification%20to%20the%20International%20Community.pdf.

General Purpose Standing Committee No 5. (May 2001). *Report on inquiry into oil spills in Sydney Harbour*. New South Wales Parliament Legislative Council. www.parliament.nsw.gov.au/lcdocs/inquiries/2205/Committee%20Report%2029%20May%202001%20-%20Inquiry%20into%20Oil%20Sp.pdf.

Guidi, G., Sliskovic, M., Violante, A. C., & Vukic, L. (September 2020). Application of the analytic hierarchy process (AHP) to select the best oil spill cleanup method in marine protected areas for calm sea condition. *Global NEST Journal*, 22(3), 354–360. https://doi.org/10.30955/gnj.002811.

Ha, M.-J. (January 2018). Modeling for the allocation of oil spill recovery capacity considering environmental and economic factors. *Marine Pollution Bulletin*, 126, 184–190. https://doi.org/10.1016/j.marpolbul.2017.11.006.

House Committee on Transportation and Infrastructure. (September 2020). *The Boeing 737 MAX report: Final committee report*. U.S. Congress.

Incident Analysis Team. (April 2000). *The response to the Laura D'Amato oil spill*. Australian Maritime Safety Authority. www.amsa.gov.au/sites/default/files/2023-11/2000-04-mp-historical-incidentlaura- damato-spill.pdf.

Keeney, R. L., & Raiffa, H. (1976). *Decisions with multiple objectives: Preferences and value tradeoffs*. Wiley.

Klein, G. A. (1993). A recognition-primed decision (RPD) model of rapid decision making. In G. A. Klein, J. Orasanu, R. Calderwood, & C. E. Zsambok (Eds.), *Decision making in action: Models and methods* (pp. 138–147). Norwood, NJ: Ablex Publishing Corporation.

Klein, G. A. (2017). *Sources of power, 20th anniversary edition: How people make decisions*. MIT Press.

Marine Incident Investigation Unit. (February 2000). *Departmental investigation the release of oil from the Italian flag motor tanker Laura D Amato at Gore Bay, Sydney Harbour on 3 August 1999*. Australian Transport Safety Bureau. www.atsb.gov.au/publications/investigation_reports/1999/mair/mair149.

National Transportation Safety Board (NTSB). (2020). Aircraft accident report: Loss of control and impact with sea lion air flight 610 Boeing 737-8 (MAX), NTSB/AAR-20/02. Washington, DC. www.ntsb.gov/investigations/AccidentReports/Reports/AAR2002.pdf?utm_source=chatgpt.com

Saaty, T. L. (1980). *The analytic hierarchy process: Planning, priority setting, resource allocation*. McGraw-Hill.

Saaty, T. L. (September 1990). How to make a decision: The analytic hierarchy process. *European Journal of Operational Research*, 48(1), 9–26. https://doi.org/10.1016/0377-2217(90)90057-i.

Sumwalt, R. L., Landsberg, B., & Homendy, J. (2019). Assumptions used in the safety assessment process and the effects of multiple alerts and indications on pilot performance. *District of Columbia: National Transportation Safety Board*. https://flightsafety.org/ntsb-urges-changes-in-aircraft-design-assessments/

U.S. Chemical Safety and Hazard Investigation Board. (2023). *Equipment fracture and fatal hydrogen chloride release at Wacker Polysilicon North America* (Investigation Report No. 2021-01-I-TN). U.S. Chemical Safety and Hazard Investigation Board. www.csb.gov/assets/1/20/wacker_polysilicon_investigation_report_final.pdf?16931=&utm_source=chatgpt.com

von Neumann, J., & Morgenstern, O. (1944). *Theory of Games and Economic Behavior*. Princeton University Press.

Weick, K. E., & Sutcliffe, K. M. (2007). *Managing the unexpected: Resilient performance in an age of uncertainty* (2nd ed., Vol. 8). Jossey-Bass.

10 Applications of Group Decision-Making – Delphi Method

10.1 INTRODUCTION

A group decision-making (GDM) is an integrated process, where participative and empathetic organizational and social culture and a team-oriented work are crucial factors in creating a group work environment. Even though GDM relates to teamwork, it requires supporting professionals to perform autonomously, explore the ability to get involved, and the appropriateness of engagements in making rational group decisions. They need to acquire intuition power to optimize their capabilities considering organizational values in the decision-making process. This mechanism succeeds unconsciously and encompasses acquired patterns of knowledge in a group environment.

Creativity is paramount in GDM that can influence individuals and organization, delivering innovative and novel ideas or concepts for task execution. Group members have social influence on each other, initiating to explore new solutions to solve problems in uncertain situations. Group decisions can be precisely made by applying social and organizational decision schemes.

Group or collaborative decision-making refers to decisions made collectively by individuals working together to achieve one or more objectives. The effectiveness and outcomes of such decisions are influenced by factors, including group size and composition, established group norms and dynamics, ethical standards within the group, organizational values, and the legislative or regulatory environment in which the organization operates.

This chapter presents the Delphi method as a structured GDM approach, detailing the process of selecting relevant experts, designing and distributing questionnaires, collecting and aggregating responses, and refining opinions through multiple rounds of feedback to support rational decision-making. It outlines the procedural steps involved in applying the Delphi method and illustrates its practical use through case studies such as the Rana Plaza collapse and the Boeing 737 Max crisis. The chapter emphasizes how professionals can leverage this method to facilitate informed, consensus-based decisions in complex and high-stakes situations.

DOI: 10.1201/9781003685449-10

10.2 GDM PROCESS

GDM is a process flow that follows sequential steps such as problem identification through brainstorming, execution of the Delphi method, creating a weighted matrix, ranking through Bayesian, Markov, or MCDM (Multi-Criteria Decision Making), assessing SWOT, designing a decision tree, and developing consensus mapping. It improves the quality of making decisions by considering available options and enhances the process by accommodating versatile knowledge from experts and their acceptance in the decision-making process. Organizations can apply GDM for their critical facilities in making decisions that can enable groups to make decisions rationally and logically. Logistical difficulties, such as scheduling and coordination, can further complicate the process. Creating an environment that promotes open dialogue and encourages constructive dissent is therefore critical to effective GDM in systems development.

The GDM is essential in systems development due to the complexity and multidisciplinary nature of complex adaptive systems. It requires the integration of diverse skills and perspectives to address technical requirements, user experience, operational constraints, and business objectives. Collaboration among team members from varied backgrounds enriches the quality of decisions, while stakeholder involvement increases acceptance and commitment to the selected solutions. By pooling knowledge, GDM enhances accountability and strengthens problem-solving capabilities. However, GDM also presents challenges, including the risk of groupthink, which can suppress alternative viewpoints and lead to poor decisions, as well as interpersonal conflict that may hinder collaboration and slow progress. In the GDM, the Delphi method plays a critical role for professionals to avoid bias and polarization that can increase cognitive efficiency, credibility, and moral reasoning through participatory and group discussions within a team.

10.2.1 OVERVIEW DELPHI METHOD

The Delphi method is a structured communication technique used in GDM that relies on a panel of experts to provide informed judgements through multiple rounds of questionnaires (Habibi et al., 2014; Hasson et al., 2025). Its primary objective is to achieve consensus while reducing bias and encouraging independent thinking. The process involves selecting relevant experts, designing and distributing questionnaires, collecting and aggregating responses, and refining opinions through successive feedback rounds. Anonymity is a key feature of the method, allowing participants to express their views freely without peer pressure, while a facilitator summarizes the results and highlights areas of agreement and important insights.

The Delphi method is applied in GDM through a series of structured and sequential steps designed to build informed consensus. First, a panel of experts with diverse backgrounds and strong domain expertise is carefully selected to ensure a broad range of perspectives. In the first round, these experts are given an initial questionnaire, often open-ended, to elicit their independent opinions, predictions, or proposed solutions to the problem at hand.

Next, a facilitator collects the responses and conducts an analysis, synthesizing the input into a concise and anonymized summary that captures key themes, points of agreement, and areas of divergence. This summary is then shared with the experts in a second round, during which they are asked to review the collective feedback and reconsider or refine their original responses.

The process may be repeated over multiple rounds to further narrow differences and stabilize responses until a reasonable level of consensus is achieved. Finally, a comprehensive report is produced that documents the final outcomes, clearly outlining areas of agreement, remaining disagreements, and the insights generated through the iterative process.

The selected GDM model for this analysis through the Delphi framework was applied within a cognition-driven decision-making context. This model is designed to systematically identify, weight, and escalate cognitive biases and risk signals within GDM processes, particularly in high-risk industrial environments.

10.2.2 Procedural Steps in Delphi Method

The Delphi method follows a few procedural steps in making decisions for professional practices, particularly for critical situations.

Step 1: Identifying Key Issues and Appointments of Experts

Organizations can appoint a multidisciplinary panel of experts with relevant knowledge like safety management, risk assessment, process engineering, project management, and human factors those who can identify and assess critical concerns for the selection process or to choose an appropriate system. Using the Delphi method, a diverse group or expert panel, including external auditors, engineers, and process systems specialists, can evaluate to deliver independent insights. The initial rounds can focus on clearly defining fundamental issues related to safety culture, risk assessment practices, and the adequacy of existing technical safety controls. They can address organizations' preferences or principles to develop key issues in the initial stage.

Step 2: Evaluation and Improvements

Once key issues are identified, experts can participate in structural and iterative discussion meetings to assess functional capacities, operational risks, and efficiency, delivering appropriate recommendations and proposing alternatives to expedite the decision-making process. This phase involves identifying potential areas such as organizational requirements, ethical values, regulations, technical features, financial strength, previous experience, and risk assessment. Through successive Delphi rounds, the expert panel could propose to choose the best option following rational decision-making, which can support an organization.

Step 3: Proposals and Recommendations

To support effective execution, a follow-up Delphi phase can focus on delivering proposals and recommendations. Experts can develop practical strategies outlining timelines, resource requirements, and accountability for each initiative. They can deliver a comprehensive proposal that guides senior management to make decisions. A continuous feedback mechanism may enable regular evaluation of implementation progress and effectiveness, addressing functional efficiencies through ongoing expert consultation.

Overall, the Delphi method enables a thorough assessment following a few sequential stages (as shown in Figure 10.1) of challenges by integrating diverse expert perspectives rather than relying solely on internal management views. Its iterative feedback process can raise risk awareness among stakeholders and foster a stronger assessment-oriented culture. By synthesizing expert knowledge, any organization can make more informed decisions regarding operational improvements and potentially address uncertainties. Furthermore, the method supports consensus-building and acceptance of appropriate measures among both employees and leadership, reinforcing shared responsibility for rational decision-making.

The Delphi method is a sequential, structured technique that supports organizations in GDM. Appointed panel of experts can contribute to engage in group work and share their valuable insights through multiple rounds of questionnaires and focus group meetings. The main goal is to receive a consensus by accommodating critical thinking from versatile knowledge experts who can avoid biases and other decision-making constraints. The method includes appointments of experts, questionnaire development, response collection through several rounds of meetings, result compilation, and feedback collection. Another critical issue is to select or appoint experts anonymously so that experts do not know each other, to make the process more credible and it can allow experts to share their opinions independently and accurately. Appointing authority can delicate a facilitator to organize meeting schedule, prepare agenda, note-taking, and meeting-minute preparation.

FIGURE 10.1 Procedural Steps of Delphi Method in Decision-Making.

10.3 USE CASE ANALYSIS – RANA PLAZA INCIDENT

10.3.1 Background

This chapter analyses the collapse of the Rana Plaza building in Bangladesh from the perspective of GDM. The disaster, which killed more than a thousand workers and left thousands injured, revealed serious flaws in leadership, regulation, and decision-making processes within the textile sector. Owners, supervisors, and authorities were affected by mental biases, economic pressures, and hierarchical structures that obstructed logical risk assessment.

The selected case study examines the collapse of the Rana Plaza building in Bangladesh on April 24, 2013. The incident is considered one of the largest industrial disasters in contemporary history. Five textile factories producing for well-known international brands were in the eight-story building. An evacuation was ordered because cracks had been observed in the structure the day before the collapse. However, the next day, the owners and supervisors forced the employees to return under threat of dismissal. The building collapsed, causing more than 2,500 people to be injured and more than 1,100 people to die (BRAC Institute of Governance and Development, 2023).

One of Bangladesh's most important economic drivers is the textile sector, which employs more than four million people, mostly women, and accounts for approximately 80% of the country's exports (EPB, 2025). However, cost reductions, intense production pressure, and lax regulation have fostered rapid growth. In this context, a work environment was created in which safety was sacrificed in favour of economic goals and decisions were made under intense external and internal pressure. Thus, the collapse of Rana Plaza exposed the ethical, organizational, and structural shortcomings of the global low-cost production model.

The fundamental purpose of the use case analysis is to demonstrate that the implementation of structured decision-making models such as the Delphi method can prevent similar tragedies. Three complementary perspectives are examined through the Delphi technique to enhance agreement and cooperation between groups; a framework for leadership and GDM focused on transparency and ethical responsibility that encourages the use of digital technologies to optimize risk response.

The starting point for the analysis is the moment when supervisors and owners chose to restart production despite noticeable cracks in the building's structure. This decision was the result of a combination of various cognitive biases and institutional errors. The company owners underestimated the danger of collapse due to overconfidence bias; employees did not refuse to follow orders due to authority bias; and inspectors, for their part, assumed that the situation was not so serious thanks to normalcy bias. Likewise, the lack of coordination between private and public actors and pressure from international brands worsened the crisis. Analysing this situation from a group, strategic, and agile perspective will provide insight into how organizations can strengthen ethics, collaboration, and responsiveness to future emergencies.

The analysis concludes by stating that the combination of collaborative work, ethical leadership, group thinking, and structured decision-making can make it possible to reduce cognitive bias, optimize communication, and increase resilience in global supply chains, particularly in high-risk industries such as textiles.

10.3.2 Purpose and Relevance

The Rana Plaza disaster in 2013 resulted from a systemic failure in GDM at various organizational levels, rather than a single technical flaw. Factory owners, engineers, government inspectors, and international buyers each possessed only partial information about structural risks and acted independently. This independence led to incoherent and unsafe collective results. The tragedy highlights how issues like poor coordination, information asymmetry, and hierarchical pressure can impair ethical decision-making in complex socio-technical systems.

To avoid failures like these, structured GDM frameworks are essential for aligning viewpoints, exchanging information, and supporting ethically responsible choices under uncertainty. The complementary method, like the Delphi technique, can provide an integrated approach for analysing and enhancing decision-making in socio-technical contexts such as Rana Plaza. In socio-technical systems engineering, making effective group decisions depends on balancing social norms, power dynamics, and ethical principles. It also involves reducing risks like groupthink, conflicts of interest, and status bias. The Delphi method makes an understanding with quantitative consensus, creating decision systems that are rational, participatory, and resistant to biases.

10.3.3 Applications of Delphi Method for Rana Plaza Incident

The Delphi Method provides a structured way to reach expert agreement via multiple rounds of anonymous feedback under conflict resolution and group consensus. This method reduces hierarchical bias and promotes reflection based on evidence. The process flow diagram followed by the Delphi method, considering the Rana Plaza incident, can be illustrated as shown in Figure 10.2.

Step 1: Problem Definition
The main decision centred on whether to evacuate or keep operations running after cracks emerged. Stakeholders faced conflicting priorities: economic survival versus human safety.

Step 2: Expert Panel Formation
A balanced panel should consist of structural engineers, labour-rights specialists, government inspectors, factory association leaders, and representatives from global brands and worker unions.

Step 3–4: Iterative Rounds and Feedback
In Round 1, experts independently assess the collapse risk using a 1–5 scale. Engineers rate it as "5 = very high," while owners facing financial pressure report "2 = low." After reviewing the anonymous aggregated results, participants revise their

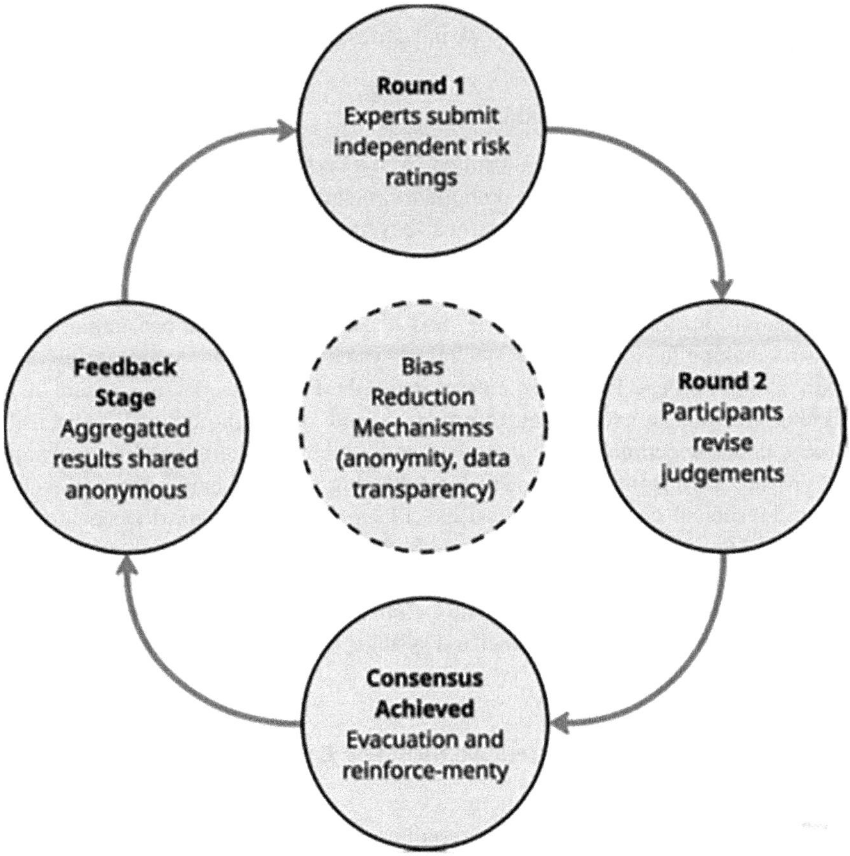

FIGURE 10.2 Process Flow Diagram for Delphi Method.

evaluations in Round 2. This feedback loop helps their judgements align towards the evidence-based recommendation: immediate evacuation and structural reinforcement before reopening.

Step 5: Implementation and Monitoring
After reaching consensus, the team collaboratively sets roles and timelines for safety implementation. Ongoing re-evaluations incorporate current information, illustrating the progressive decision-making model, which emphasizes continuous learning and adaptive reasoning. The Delphi method thereby implements substantive rationality, assessing actions by their expected results, while reducing authority bias, anchoring, and groupthink through the anonymity and transparency of data.

10.3.4 Decision-Making Insights

The use case analysis addresses the cognitive and ethical issues about heuristics, biases, and group thinking. At Rana Plaza, professionals' optimism bias and confirmation bias led them to misinterpret the lack of immediate collapse as a sign of safety. Their focus on production deadlines further clouded their judgement. The Delphi technique could help them to address these issues by encouraging reflection through structured feedback.

GDM refers to how organizational leaders guide long-term direction through group thinking and critical choices. The rational decision-making framework and situational theories emphasize that organization can execute GDM analysis with intuition. For example, the board of Directors notes that strategic decisions "focus on the medium- to long-term future direction of the business" and require balancing short and long-term goals with the organization's vision. Leadership styles also matter; studies show situational leaders make more adaptive strategic decisions, and transformational leaders can foster the innovation needed to achieve strategic aims through GDM processes.

In practice, GDM involves a clear process: experts gather information, define and identify problems, consult and deliver opinions, and evaluate alternatives to make a choice that advances long-term organizational objectives. Practical GDM thus combines analytical rigour with ethical and collaborative judgement, mitigating cognitive biases and considering stakeholder welfare.

The Rana Plaza disaster highlights the deadly risks posed by rigid hierarchies and restricted communication. By institutionalizing distributed authority, iterative feedback, and digital transparency, organizations can identify and respond to early warning signs before crises escalate. Achieving resilience, ethical accountability, and sustainable safety across high-risk sectors, ranging from apparel manufacturing to construction and mining, requires a comprehensive cultural transformation framework. The Delphi technique effectively addresses such challenges by converting diverse expert judgements into evidence-based consensus while maintaining anonymity, thereby reducing the influence of hierarchical power dynamics.

Applied to Rana Plaza, this stack would have legitimized front-line dissent via anonymous Delphi rounds, bound leadership to pre-declared stop thresholds, created verifiable evidence trails that brands and regulators could not ignore, and compressed time from detection to evacuation – yielding fewer false "all-clear" decisions, faster remediation, and more durable trust.

10.4 USE CASE ANALYSIS – BOEING 737 MAX

10.4.1 Background

The Boeing 737 MAX project is one of the most representative complex decision-making failures in contemporary aviation engineering management (JATR, 2019). To compete with the Airbus A320neo, Boeing redesigned the existing 737 platform and installed larger and more energy-efficient engines, thereby changing the aircraft's aerodynamic balance. To compensate for this change, engineers developed the Manoeuvring Characteristics Augmentation System (MCAS) – a programme that automatically adjusts the pitch of the nose to maintain a similar handling feel to earlier models.

However, under intense market competition and urgent delivery pressure, the organization's focus has shifted from strict security verification to rapid evidence collection and cost control. Boeing has classified MCAS as a "minor" or "major" safety system, rather than a "critical" system, thereby avoiding redundant sensor configurations and stricter pilot training requirements. Although this classification decision saves time and costs, it leaves hidden dangers for single point failures. As a failed angle of attack sensor repeatedly triggered MCAS, it resulted in two fatal air crashes. This chapter reveals how time pressure, information silos, and herd behaviour collectively distort rational decision-making in high-risk, cross-departmental environments, even in organizations with strong technological capabilities.

10.4.2 Purpose and Relevance

The purpose of this use case analysis is not simply to explore mechanical failures, but to analyse how GDM frameworks affect organizations' judgements and choices in high-risk situations. The analysis focuses on the cognitive biases, organizational culture, and procedural deficiencies presented by Boeing in the MCAS design and certification process.

By introducing a complementary decision-making framework following the GDM, the use case analysis aims to demonstrate a multidimensional rational improvement path. The framework explains distinct levels of decision-making logics that can reduce status bias and groupthink through structured expert opinion collection.

The core scenario analysed in this analysis is Boeing's decision on the risk classification and governance of the MCAS system during the 737 MAX certification phase. At that time, companies were caught in a dilemma between commercial urgency and security. If MCAS is identified as a "critical flight control system," it would require the addition of multiple sensor redundancies, stricter software validation, and pilot training, which could inevitably lead to project delays and increased costs; if handled according to the "minor" or "major" system, progress and competitiveness can be maintained, but there is a significant safety hazard of single point failure.

This classification decision was heavily influenced by both organizational communication breakdowns and cognitive biases. While the engineering team recognized the potential risks, they hesitated to question management's overly optimistic assumptions. At the same time, regulatory agencies placed excessive reliance on manufacturers' self-certification, fostering a misleading sense of procedural compliance. As a result, groupthink prevailed, allowing short-term performance objectives to take precedence over long-term safety obligations.

Therefore, this chapter includes the MCAS case considering a decision-making environment that combines technical complexity, political sensitivity, and ethical consequences. In the absence of a structured and transparent decision-making framework, organizations do not make a single mistake but rather fall into systematic errors.

10.4.3 Applications of Delphi Method for MCAS Incident

The Delphi method was developed in the 1950s by the Air Force-sponsored RAND Corporation as an effective way to obtain the most reliable consensus among a group of experts. According to Okoli Pawlowski (2004), the Delphi method can be described as an iterative tool for structuring a group communication process and therefore work together on a complex problem by providing feedback on individual contributions, opportunities to revise ideas, anonymity for individual responses as well as some evaluation of group judgement. In practice, the Delphi method operates through a series of anonymous, iterative rounds in which experts independently provide judgement, receive controlled feedback on the group's responses, and are encouraged to reconsider their views considering the collective information, thereby generating a reliable consensus and reducing the influence of group effect or dominant individuals.

The Delphi method is currently widely used in various GDM processes, such as forecasting future trends in technology or policy, risk assessment, and strategic planning in government, business and research, particularly in new product development. Systematic and interactive nature of the Delphi method is crucial for developing and evaluating future scenarios, considering complexities and unforeseen circumstances. Similarly, in the field of aerospace engineering, the structured Delphi method has been shown to be widely used in engineering design in the aviation field and to guide product planning under high complexity and uncertainty. Therefore, Delphi, as a structured GDM tool in the technical field of interdisciplinary consensus, is highly relevant to Boeing's engineering and safety assessment.

10.4.4 Rationality for Delphi Method in MCAS Incident

The decision-making situation surrounding the MCAS between 2016 and 2018 required Boeing engineers to balance multiple technical and safety factors, including sensor redundancy, control authority, human–machine interface (HMI) design, and hazard classification, under considerable time and cost pressures. Choices regarding the number of sensor inputs, control authority, and hazard classification had to be made quickly, often based on assumptions about rapid pilot recognition and correction. These circumstances created a high-risk, technically complex, and cross-functional situation that required structured expert judgement across multiple disciplines, making it well-suited for structured team decision-making methods like Delphi.

From a psychological perspective, the design and evaluation process of MCAS reflected typical characteristics of groupthink. Groupthink is a psychological phenomenon that is particularly prevalent in GDM. Its characteristics include overconfidence in the group, collective rationalization of past decisions, and the shared illusion of consensus that judgements conform to the majority view. In the case of MCAS, the pressure of groupthink limited the expression of dissenting opinions. The engineering and airworthiness teams appeared to underestimate findings indicating single-point failure risks and the need for human factors testing, while favouring evidence supporting MCAS as a moderate enhancement (JATR, 2019). Furthermore, communication gaps further increased the potential for erroneous decisions. Due to poor cross-departmental communication, the safety team's opinions were not adequately reported. Furthermore, due to overconfidence and the illusion of group rationalization, the team ignored evidence of insufficient training and did not take the safety team's opinions seriously.

The Delphi method directly reduces the occurrence of these phenomena through anonymity, iterative feedback and controlled communication, allowing the moderator to gather experts' individual brainstorming, summarize high-quality multi-perspective GDM results, and make a more balanced assessment of complex safety scenarios.

The primary goals of employing the Delphi methodology were to improve the accuracy of Boeing's engineering and safety decisions regarding the MCAS system, reduce the influence of groupthink in GDM, and achieve a structured expert consensus on system design trade-offs in a high-pressure production environment.

10.4.5 Representation of the Delphi Process

The Delphi process for the MCAS case consists of six iterative steps, with each round anonymizing responses, summarizing reasoning, and providing controlled feedback for reflective revision – the process flowchart can be illustrated as shown in Figure 10.3.

10.4.6 Decision-Making Insights

The application of the Delphi method provides several useful insights for improving engineering decision-making in Boeing's MCAS project. Using anonymity, experts from diverse fields expressed honest opinions, helping to identify potential risks early. Furthermore, disagreements among experts were emphasized, revealing overlooked issues. Furthermore, the Delphi process records each round of feedback, allowing Boeing and other airlines to review the GDM process and improve future decisions.

10.5 CONCLUSION

The analysis of the Rana Plaza incident underscores the critical value of structured GDM in high-risk, complex environments. The Delphi method, through anonymous and iterative rounds of feedback, allows organizations to generate consensus-driven decisions even under conditions of incomplete information and fragmented authority. By separating technical judgement from personal influence, it mitigates the risks of groupthink and enables more balanced, robust strategic decisions. When applied systematically, this approach can legitimize dissent, enforce timely evacuation, and create verifiable records that ensure accountability, reducing the likelihood of false "all-clear" decisions and accelerating remediation.

However, the effectiveness of such methods is shaped by practical constraints, including uneven technological readiness, limited digital literacy among lower-tier suppliers, entrenched authoritarian cultures, and the costs of digital monitoring. These challenges can be addressed by phasing technology adoption, coupling tools with training, codifying clear trigger ladders, distributing stop-work authority, employing independent facilitation for Delphi processes, and requiring third-party verification before operations resume. From a policy perspective, organizations and

FIGURE 10.3 Six-Step Delphi Process Flowchart for MCAS Safety Evaluation.

regulators should institutionalize evacuation triggers, archival rationales for decisions, tamper-evident inspection records, and metrics tracking detection-to-action latency. Collectively, these practices embed a durable duty of care, ensuring that ethical and safety responsibilities are systematically upheld across the supply chain while translating analytical insights into enforceable operational routines.

BIBLIOGRAPHY

Ali, O., Murray, P. A., Muhammed, S., Dwivedi, Y. K., & Rashiti, S. (2022). Evaluating organizational level IT innovation adoption factors among global firms. *Journal of Innovation & Knowledge*, 7(3), 100213. https://doi.org/10.1016/j.jik.2022.100213.

Bossavie, L., Cho, Y., & Heath, R. (2023). The effects of international scrutiny on manufacturing workers: Evidence from the Rana Plaza collapse in Bangladesh. *Journal of Development Economics*, 163, 103107. https://doi.org/10.1016/j.jdeveco.2023.103107.

BRAC Institute of Governance and Development. (2023). *Ten years after Rana Plaza: Remembering, learning, and moving forward*. BRAC University.

Carmody-Bubb, M. (2023). *Cognition and decision making in complex adaptive systems: The human factor in organizational performance* (1st ed.). Springer International Publishing. https://doi.org/10.1007/978-3-031-31929-7.

Checkland, P. B. (1989). Soft systems methodology. *Human Systems Management*, 8(4), 273–289. https://doi.org/10.3233/hsm-1989-8405.

Clean Clothes Campaign. (n.d.). Rana Plaza. https://cleanclothes.org/campaigns/past/ranaplaza.

Dalkey, N., & Helmer, O. (1963). An experimental application of the DELPHI method to the use of experts. *Management Science*, 9(3), 458–467. https://doi.org/10.1287/mnsc.9.3.458.

Export Promotion Bureau (EPB). (2025). *Export performance of Bangladesh (FY 2023–2024 / 2024 calendar estimates)*. Government of Bangladesh. https://epb.gov.bd/site/view/epb_export_data.

Guillemot, J., Aoussat, A., Corvez, A., & Boudard, J. (2024). Approach for participatory Delphi method for structuring voices, population empowerment and research training. *MethodsX*, 13, 103019. https://doi.org/10.1016/j.mex.2024.103019.

Habibi, A., Sarafrazi, A., & Izadyar, S. (2014). Delphi technique theoretical framework in qualitative research. *The International Journal of Engineering and Science*, 3(4), 8–13. https://www.researchgate.net/publication/272177606_Delphi_Technique_Theoretical_Framework_in_Qualitative_Research.

Hasson, F., Keeney, S., & McKenna, H. (2025). Revisiting the Delphi technique – Research thinking and practice: A discussion paper. *International Journal of Nursing Studies*. 168, 105119. https://doi.org/10.1016/j.ijnurstu.2025.105119

Human Rights Watch. (2015). "Whoever raises their head suffers the most": Workers' rights in Bangladesh's garment factories. https://features.hrw.org/features/HRW_2015_reports/Bangladesh_Garment_Factories/index.html.

Islam, M. N., Hossain, M. M., & Ornob, M. S. S. (2024). Business research on Industry 4.0: A systematic review using topic modelling approach. *Future Business Journal*, 10(1), 111. https://doi.org/10.1186/s43093-024-00398-2.

Jaafar, M., Khan, K. N., & Salman, A. (2025). A systematic review and framework for organizational agility antecedents towards Industry 4.0. *Management Review Quarterly*. 76(1), 487–512. https://doi.org/10.1007/s11301-025-00489-6.

Joint Authorities Technical Review (JATR). (2019). *Boeing 737 MAX flight control system: Observations, findings, and recommendations*. Federal Aviation Administration.

Kameda, T., & Sugimori, S. (1993). Psychological entrapment in group decision making: An assigned decision rule and a groupthink phenomenon. *Journal of Personality and Social Psychology*, 65(2), 282–292. https://doi.org/10.1037/0022-3514.65.2.282.

Magistretti, S., & Trabucchi, D. (2025). Agile-as-a-tool and agile-as-a-culture: A comprehensive review of agile approaches adopting contingency and configuration theories. *Review of Managerial Science*, 19, 223–253. https://doi.org/10.1007/s11846-024-00745-1.

Mrugalska, B., & Ahmed, J. (2021). Organizational agility in Industry 4.0: A systematic literature review. *Sustainability*, 13(15), 8272. https://doi.org/10.3390/su13158272.

Ndou, V., Ingrosso, A., & Di Girolamo, A. (2024). Framework for agile transformation: Guiding organizations through cultural, structural, and competency shifts in project management. *Administrative Sciences*, 14(11), 301. https://doi.org/10.3390/admsci14110301.

Nguyen, T., Le, C. V., Nguyen, M., Nguyen, G., Lien, T. T. H., & Nguyen, O. (2025). The organisational impact of agility: A systematic literature review. *Management Review Quarterly*, 75, 2709–2757. https://doi.org/10.1007/s11301-024-00446-9.

Okoli, C., & Pawlowski, S. D. (2004). The Delphi method as a research tool: An example, design considerations and applications. *Information & Management*, 42(1), 15–29. https://doi.org/10.1016/j.im.2003.11.002.

Reguant-Alvarez, M., Alvarez Torrado-Fonseca, M., & Fonseca, M. (2016). El metodo Delphi. *REIRE*, 9, 1916. https://doi.org/10.1344/reire2016.9.1916.

Sigala, A., & Langhals, B. (2020). Applications of unmanned aerial systems (UAS): A delphi study projecting future UAS missions and relevant challenges. *Drones*, 4(1), 8. https://doi.org/10.3390/drones4010008.

Snowden, D. J., & Boone, M. E. (2007). A leader's framework for decision making. *Harvard Business Review*, 85(11), 68.

Toločka, E., & Maceika, A. (2024). Delphi method application for evaluation of innovative product development scenario. *Entrepreneurship and Sustainability Issues*, 11(3), 37–49. https://doi.org/10.9770/jesi.2024.11.3(3).

U.S. House Committee on Transportation and Infrastructure. (2020). The design, development & certification of the Boeing 737 MAX. Final Committee Report. https://democrats-transportation.house.gov/committee-activity/boeing-737-max-investigation?utm_source=chatgpt.com

11 Leadership and Strategic Decision-Making – Cynefin Framework

11.1 INTRODUCTION

Critical issues in uncertain situations can cause problems in different contexts, where leaders demand contextual information regardless of complexity. Professionals or senior management can be stuck in entrained thinking, where past experiences, practical knowledge, training, structured decision-making tools, and situational awareness enhance their capabilities to innovate new ideas. Leaders can become satisfied when things go well, causing them to miss changes due to overconfidence and get much time to react to critical situations. Best benchmarking practices can be useful but blindly following them can be challenging if unknown events occur. Complex situations have several right options and are known as "known unknowns" combination and leaders must have the capacity to tackle such situations by applying their sense, analytical knowledge, and proactive responses.

Effective leaders must be strategic, decisive, communicative, and curious for organizational success. Professionals can create a framework using past and future scenarios to understand their organization's context better. Viewing the world in terms of complexity is crucial beyond mathematical models. Leaders, decision-makers, and team members can gather information for an efficient command-and-control approach. They can prepare concise directives and deliver rational decisions seamlessly and follow derivative processes for optimal outcomes. Ensuring adherence to best practices or restructuring processes can maximize efficiency while minimizing communication needs in cases of limited disagreement.

Professionals, even with highly developed expertise, may inadvertently overlook innovative ideas from non-experts, potentially missing critical insights or opportunities for creative solutions. To address this, leaders should actively engage domain experts while remaining open to unconventional perspectives and leveraging advanced analytical tools. At the same time, they must avoid "analysis paralysis," where over-deliberation, limited viewpoints, or professional egos hinder the consideration of novel concepts. In complex or uncertain situations, effective decision-making requires carefully balancing the pursuit of the optimal solution with the need to act rationally, particularly when data are incomplete or circumstances are uneven.

 DOI: 10.1201/9781003685449-11

Major changes in uncertain situations lead to arbitrariness that can cause organizational instability and lead to complex decision-making. Leaders or senior management can conduct experiments and apply leadership and strategic decision-making methods to reveal new patterns, sensible ideas, innovative solutions, and situational responses. For example, initially YouTube didn't include the feature for accommodating streaming video that exists today. However, as services demanded more from the provider, YouTube addressed and developed new features on their platforms. The platform not only incorporated new technologies but also adopted flagging and content policies to guide hosts and users to maintain political views and expressions and sponsors a debate for presidential hopefuls with video feeds from the site.

Leadership and strategic viewpoints in decision-making based on crisis management have constraints. Emergency situations are infrequent, so professionals are not always ready to make decisions in such critical situations, and this uncertainty is applicable to leadership practices. Leaders who handle crises may not have the capability to make instant decisions by considering contextual information and past experiences. There are a few established methods that can relieve them from these uncertainties. Successful leaders may show their capabilities to solve identified problems in chaotic contexts, proving themselves as exceptional in strategic leadership. Chaotic contexts can inflate leaders' egos and make them confused, impeding their ability to perform effectively and making it difficult to make accurate decisions. To ensure rational decision-making during chaotic situations, professionals can follow leadership and strategic decision-making processes to act promptly in uncertainty.

Moreover, highly successful leaders in chaotic and complex contexts can develop an overinflated self-image, becoming legends in their own minds. When they generate cult-like adoration, leading actually becomes harder for them because a circle of admiring supporters cuts them off from accurate information. Chaotic and complex contexts can inflate leaders' egos, hindering their ability to lead effectively. Admirers can shield them from accurate feedback, making it hard to receive the information they need to make good decisions. Leaders can apply leadership and strategic decision-making methods in uncertain situations by applying robust analytical techniques and can focus on identifying probable solutions.

Effective leaders are malleable and equip their teams to understand diverse contexts. They correct their behaviours according to the context of their work. However, few leaders prepare their organizations to handle critical situations in different contexts. Leaders must delicate authority, be confident in their judgment, be flexible to change, and embrace complexity to address uncertainties. To be an effective leader, one must possess critical thinking skills, the ability to solve problems, and the ability to make informed decisions in real time. Strong deductive reasoning is crucial for effective project planning. Inductive reasoning is essential for leaders to make timely decisions in uncertain and risky situations. It enables them to function well with incomplete data, assess risks, develop contingency plans, adapt to changes, and determine when to move forward or pivot.

Professionals are committed to attaining professional excellence and fulfilling the organization's vision with strong determination. They can make rational decisions with strong leadership and strategic viewpoints by applying critical thinking and considering contextual perspectives. Leaders are those who welcome and adapt to change while excelling in fast-paced environments and make a strategy to fulfil organizations' goals. They can be proactive and use logical arguments and apply established methods to address issues. Effective decision-making encompasses effective solutions for complex problems, exploring multiple alternatives, and critically assessing past experience to correlate with current requirements. Leaders also respect workplace culture and deliver rational decisions to achieve the organizations' vision and mission. They are capable of inspiring their team, moving it forward with enthusiasm and encouragement, and motivating team members to deliver their best. They limit social engagements to work-related purposes and act as a professional and focused-oriented approach in workplace relationships with colleagues.

11.2 LEADERSHIP IN DECISION-MAKING

Great leaders balance emotions with reason, make tough decisions, and have a positive impact on people and organizations. This involves managing change, uncertainty, stress, and negative reactions. A skilled leader should validate potential solutions using data, combine knowledge and intuition to find the best solution, assess ideas and outcomes, and – make timely decisions.

Effective leadership decision-making relies on three essential skills that support rational choices. First, logical reasoning enables leaders to analyse facts thoroughly, evaluate pros and cons, and maintain self-awareness to guide sound decisions. Second, problem-solving allows leaders to navigate complex business or talent-related challenges, considering multiple variables and perspectives to identify the best course of action. Third, emotional intelligence equips leaders to manage their emotions, communicate effectively, and maintain composure during high-pressure decision-making. Leaders should prioritize urgent or stressful decisions to prevent procrastination and reduce stress, gather comprehensive information to ensure decisions are well-informed, and avoid making choices based on emotion alone, waiting instead for evidence and facts to guide their actions.

Leaders must make rational and judgmental decisions, and effective leadership decision-making delivers positive outcomes such as making well-informed decisions to address organizational challenges, ensuring positive outcomes and uninterrupted business operations. Good leaders inspire productivity through informed decision-making, set an example for their team, and foster communication, collaboration, and engagement. Effective decision-making can save time and provide clear direction. It reduces errors, aids delegation, and eliminates confusion. Quick and effective decision-making boosts productivity by minimizing delays and inefficiencies, allowing employees to start working faster.

Professionals in leadership roles should adopt a holistic approach that integrates leadership skills with strategic impact, carefully balancing vulnerabilities and potential outcomes. They need to think broadly, anticipate negative consequences, and assess both the likelihood and severity of risks. Effective decision-making weighs short- and long-term value, avoiding narrow short-term thinking in favour of sustainable strategies that account for immediate and future costs and benefits. To maximize impact, leaders must balance present needs with long-term objectives. Clear communication with stakeholders is essential – explaining the rationale and substance of decisions without seeking approval or consensus. Timeliness is equally critical: identify core issues, address them promptly, and continuously evaluate outcomes to refine judgment and grow as a leader.

11.3 STRATEGIC DECISION-MAKING – CYNEFIN FRAMEWORK

11.3.1 Background

Strategic decision-making is critical for professionals to uphold leadership, particularly in uncertain and complex situations. Following only previous rules in new situations can deliver wrong decisions that create uneven consequences for organizational growth. Leaders can apply methodological approaches to assess the situation and identify strategic viewpoints before making decisions. Professionals can adopt a comprehensive strategy to address uncertainties and complex situations. Uncertainties in complex adaptive systems are random, but their decision-making can be confirmed through rational, judgmental, and methodical mechanisms. Cynefin framework, an efficient method, was introduced for strategic decision-making to address complex and uncertain situations (van Beurden et al., 2013). The Cynefin is a Welsh term meaning "habitat." It provides a framework for analysing uncertainty in a different way rather than offering traditional methods or decision-making processes.

The Cynefin framework, developed by Dave Snowden, is a versatile model for understanding and navigating complexity, guiding leaders to adapt their strategies in uncertain and dynamic situations (Snowden & Boone, 2007). It categorizes decision-making contexts into five distinct domains: obvious (Simple) situations, where clear cause-and-effect relationships allow established best practices to be applied effectively; complicated contexts, which require expert analysis to uncover causal links; complex environments, characterized by interdependent elements and unpredictable outcomes, where solutions emerge through experimentation and iterative learning; chaotic situations, where no clear patterns exist and immediate, decisive action is necessary before analysis can take place; and disorder, an ambiguous state in which the nature of the problem is unclear, requiring decomposition into the other domains to enable effective decision-making.

The framework was introduced in 2007 as a strategic decision-making method to address critical situations. The Cynefin is grounded in sense-making theory and the study of complex adaptive systems, where decision-makers must interpret signals and adjust actions dynamically based on feedback. Unlike linear or rational-choice models, Cynefin acknowledges bounded rationality, the idea that leaders cannot know all variables during high-stress incidents. As such, it encourages context-aware leadership, where decisions are matched to the environment's nature rather than imposed via a one-size-fits-all approach. This reflects modern strategic leadership theory, which emphasizes flexibility, stakeholder responsiveness, and systems thinking in volatile environments.

The Cynefin framework (pronounced kuh-NEV-in) categorizes decision environments into five domains. Initially, it supports leaders in understanding the variable nature of the complex situation and making rational decisions based on critical circumstances. Later, Snowden and Boone identified two domains: simple and complicated. Simple domains have clear cause–effect relationships and best practices, while complicated domains have multiple solutions, requiring experts to analyse before responding. Decision-making may take more time. It is widely used in leadership strategy, risk governance, and engineering safety management, particularly where uncertainty and system dynamics are involved.

The Cynefin framework, developed by Snowden and further refined by scholars such as Nachbagauer, enables leaders to identify the type of environment they are operating in and determine appropriate responses (Nachbagauer, 2021). The framework categorizes contexts into five domains, each varying in levels of order, flexibility, and predictability. As Nachbagauer notes, ordered domains can be analysed to forecast outcomes, while unordered domains require decisive action to stabilize conditions because results are uncertain and often unpredictable. Failure to adapt leadership approaches to shifting circumstances can lead to serious consequences under uncertainty. Applying the Cynefin framework retrospectively to past scenarios provides valuable insights into how adaptive leadership can recognize emerging issues and implement actions that are appropriately aligned with the complexity of the situation.

The Cynefin framework facilitates professionals in decision-making in complex systems. It can assist leaders classify decision-making contexts, recognize appropriate processes, and follow a methodological approach to respond to unwanted situations. The framework is derived from complexity science and provides a lens to understand multi-causal dynamics in contexts, organizations, and communities.

11.3.2 Four Domains of Cynefin Framework

The Cynefin framework delineates four domains: Clear (Simple), Complicated, Complex, and Chaotic, which concentrate on a central "Confusion or Disorder" area when the context is unclear. Each domain requires a different decision-making approach. In Clear and Complicated (ordered) domains, cause-and-effect relationships are known or can be discovered, so leaders use best practices or expert analysis. In Complex and Chaotic (unordered) domains, uncertainty prevails, requiring experimentation or urgent action.

When we deal with Clear situations, problems and their connections are easily understood and usual practices are effective. Then, problems in the Complicated domain are analysed in detail by experts and have reliable outcomes once the proper method is applied. In Complex domain, it is only after events unfold that the connections between them can be understood. However, because there is no clear connection between causes and effects in the Chaotic domain, actions must be taken right away to help stabilize everything. Lastly, the Disorder occurs when it's not certain which of the other four domains is the best match.

The four domains of Cynefin decision-making, such as predictable, unpredictable, complicated, and chaotic, guide professionals to categorize the situational impacts. The first two can be managed using normative decision strategies, while the latter requires naturalistic decision strategies. These domains focus on extracting information and leading a team to make strategic decisions.

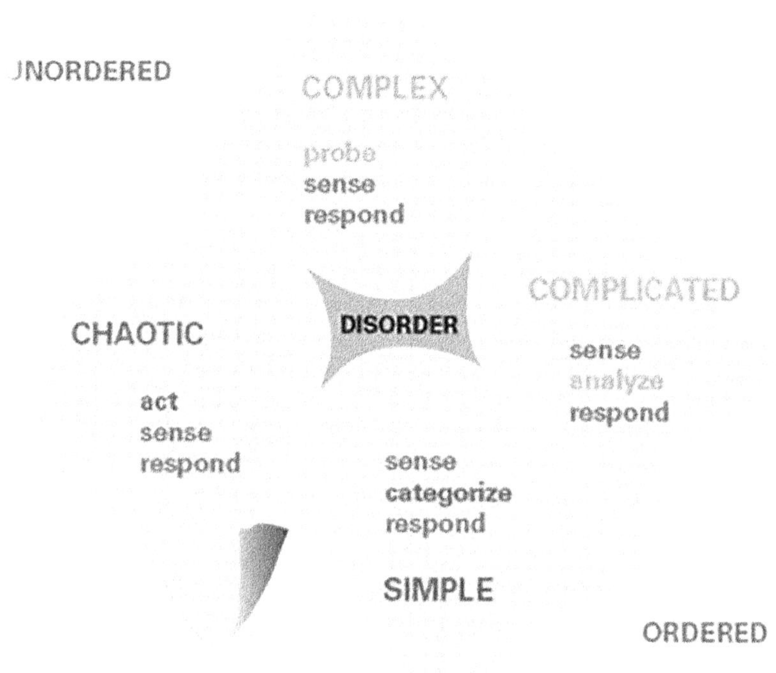

FIGURE 11.1 Four Domains of Cynefin Framework.

Problem-solving in complex environments requires a mental representation of the environment's relational structure. Creative approaches are needed to find a solution that is not easily discovered. Leaders should be wary of small changes that can cause big shifts in complex systems. They must remain open-minded, probe, sense, and respond to avoid bias. It's important to remember that emerging domains may lack established best practices. The four domains of Cynefin framework is shown in Figure 11.1.

In a complex domain, leaders must be patient and coordinate with their teams to better assess situations. They should also avoid taking control too soon, which can decrease subordinates' input.

Chaotic domains are rare and short-lived, involving shifting cause-effect and time pressure, making finding the right answer difficult. Leaders in this domain must act to gain control, assess the environment, and respond accordingly. Strategic decision-making involves movement between chaotic and complex domains, but most decisions involve complex and complicated domains. In chaos, leaders must quickly establish order and move towards complexity. The team can then identify patterns and solutions through trial-and-error and mental simulations. Teamwork is crucial. Success breeds complacency and often leads to a collapse into chaos.

The disorder domain lacks leadership and teamwork, leading to arguments and confusion among multiple perspectives. Effective communication and coordination among various teams involved in a decision domain, especially in complex and chaotic scenarios, are crucial to prevent disorders. Distributed decision-making is the practice of making coordinated decisions while geographically dispersed, as exemplified by Effective Distributed Decision Making (EDDM). Communication is critical in making effective decisions in remote team decision-making due to the lack of co-location among key roles and players.

This chapter extends the practical implications of the Cynefin framework through critical analysis of three distinct use case scenarios. The aim of this use case analysis is to demonstrate how a strategic decision-making approach like the Cynefin framework can support complex decision environments by identifying and classifying domains of Cynefin, reducing cognitive bias, and improving outcome defensibility. The case study serves as a practical scenario for applying this model, illustrating its effectiveness in both pre-crisis planning and real-time incident response.

Professionals and senior management often use the Cynefin framework to analyse strategic decisions in high-stakes situations characterized by uncertainty, due to its capacity to guide organizations through complexity. Unlike traditional linear decision-making models, which assume problems have clear and predictable solutions, the Cynefin framework recognizes that certain contexts are too dynamic and interconnected for conventional approaches to be effective.

11.4 USE CASE ANALYSIS

11.4.1 Scenario – Chernobyl Nuclear Explosion

The Chernobyl nuclear disaster, occurring on April 26, 1986, stands as a stark and enduring example of the catastrophic fallout from human error, design flaws, and systemic organizational failures (Bennett et al., 2006). Situated near Pripyat in the Ukrainian SSR, the Chernobyl Nuclear Power Plant was a vital energy source for the Soviet Union (currently named as Russia), housing four RBMK-1000 graphite-moderated reactors. The event itself was a catastrophic power excursion during a routine safety test on Reactor No. 4, leading to a series of violent explosions that breached the reactor core and unleashed a significant plume of radioactive material into the atmosphere.

The immediate consequences were devastating, with numerous fatalities among plant workers and firefighters due to acute radiation syndrome, and the rapid evacuation of over 116,000 people from the surrounding exclusion zone. In the long term, Chernobyl has resulted in widespread environmental contamination across parts of Europe, a demonstrable increase in thyroid cancer rates, and immense socio-economic burdens on the affected regions (Krukovskyi et al., 2018). This disaster fundamentally reshaped global perceptions of nuclear power safety and crisis management, forcing a re-evaluation of design, operational protocols, and international cooperation.

The initial decision-making context at Chernobyl was characterized by extreme complexity and chaos. To properly understand the varied nature of the decision situations faced, we can introduce the Cynefin framework. This powerful sense-making tool, developed by David Snowden, helps categorize environments into five domains: Simple (known knowns), Complicated (known unknowns), Complex (unknown unknowns), and Chaotic (unknowable), with Disorder in the middle. Immediately following the explosion, the situation was largely chaotic, demanding immediate action without a clear understanding of the full scope or appropriate response. As more information emerged, it evolved into a complex process, requiring iterative experimentation, and later became complicated as specific problems like reactor cooling and decontamination could be addressed with expert analysis. Over time, elements of long-term management moved into the simple domain as best practices were established.

By applying the Cynefin framework, we can better understand how misalignment between the actual situation and the chosen decision-making approach contributed to the escalation of the disaster. The primary aim of this use case analysis is to leverage the Chernobyl case study to analyse and identify effective decision-making strategies for complex organizational challenges. This chapter explores how various decision-making models – encompassing aspects of leadership, strategic thinking, innovation, and agility – could have been, or were, applied to tackle the identified issues. By examining Chernobyl through this lens, the use case analysis draws crucial lessons that directly link to contemporary organizational decision-making challenges, highlighting the critical need for robust, adaptive, and strategic frameworks in times of crisis.

The Cynefin framework highlights why varied leadership approaches were essential for tackling Chernobyl's diverse long-term strategic issues. The framework, as shown in Figure 11.2, could assist leaders in categorizing strategic challenges to guide appropriate responses:

I. Complicated (Analyse–Sense–Respond): These situations have knowable unknowns where expert analysis can identify cause-and-effect. For example: The engineering task of designing and constructing the New Safe Confinement (NSC) over the original deteriorating sarcophagus. This was a complex technical problem solvable through rigorous engineering analysis, detailed planning, and expert execution. Developing precise decontamination protocols for specific areas also fits this domain.
II. Complex (Probe–Sense–Respond): These situations feature unknowable unknowns where solutions emerge through experimentation and learning. For example: Managing the social reintegration of hundreds of thousands of displaced people or addressing the long-term psychological trauma from radiation exposure and displacements. There were no existing blueprints, requiring iterative social programs, mental health support, and adaptive strategies based on community feedback and evolving needs.

The Cynefin framework was chosen for its applicability to complex, uncertain environments like the Chernobyl nuclear explosion. Unlike traditional linear models (e.g., Rational Planning Model), Cynefin accounts for emergence, ambiguity, and nonlinearity, all of which were present in this case. Its emphasis on recognizing

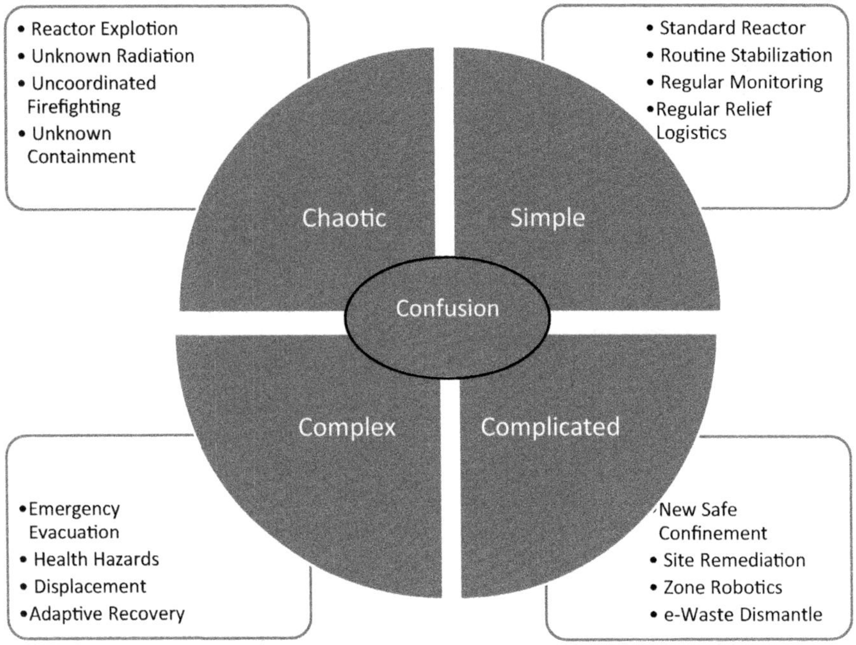

FIGURE 11.2 Chernobyl Decision-Making: A Cynefin Framework.

domain boundaries and adapting leadership style is ideal for analysing engineering decisions that span ethics, risk, and team coordination.

11.4.2 Scenario – Australian Census Website Outage

The 2016 Australian Census website failure exposed critical weaknesses in cybersecurity preparedness and crisis response (ABS, 2017). This chapter evaluates how advanced strategic decision-making models like the Cynefin framework can improve strategic responses to digital infrastructure threats. Drawing on the framework, the analysis demonstrates how structured and adaptive methodologies can mitigate the impact of uncertainty and stakeholder pressure in complex environments. The findings highlight the importance of early risk identification, structured expert consensus, and agile crisis leadership, highlighting the demand for adopting the framework for informed and resilient decision-making under digital threat conditions.

The increasing complexity of digital infrastructure and the growing prevalence of cybersecurity threats have underscored the need for robust, agile, and evidence-based decision-making frameworks. This chapter addresses one such critical decision context: the cybersecurity failure during the 2016 Australian Census. On Census night, a coordinated Distributed Denial of Service (DDoS) attack, combined with system vulnerabilities and poor communication, led to the shutdown of the Australian Bureau of Statistics (ABS) website. This incident provides a compelling backdrop for exploring how advanced strategic decision-making models can improve preparedness, response, and public trust in high-stakes digital environments.

The 2016 Census website failure highlights a scenario characterized by uncertainty, urgency, and multifaceted risks (JCPAA, 2017). The ABS was tasked with managing public participation in a national data collection effort while safeguarding sensitive information in an online-only format for the first time. The decision to take the system offline during the incident was reactive and lacked structured justification, revealing limitations in leadership foresight, stakeholder engagement, and contextual awareness.

The 2016 Australian Census website outage presents a compelling case for applying the Cynefin framework to understand how leadership decisions could evolve across domains of complexity. The incident, triggered by DDoS attacks, infrastructure failures, and inadequate crisis planning, demanded adaptive and context-aware leadership. The Cynefin framework offers a strategic decision-making model that helps leaders identify the nature of a situation and tailor their actions accordingly. The outage transitioned through multiple domains, including complicated, chaotic, complex, and eventually clear, highlighting the need for fluid leadership capable of shifting decision styles as the crisis unfolded.

Through this scenario, the analysis explores decision-making across distinct phases – pre-crisis, crisis, and post-crisis – each demanding different cognitive approaches and strategic tools. By applying the Cynefin framework alongside group decision-making techniques, the analysis evaluates how leadership could have adapted decision styles as the situation evolved from a complicated to a chaotic domain as shown in Figure 11.3. The subsequent integration of other decision-making tools can further support structured prioritization of risk mitigation strategies during the recovery phase. The four domains of the Cynefin framework can be explained for the Australian census outage incident as below.

11.4.2.1 Complicated Domain: Pre-Crisis Infrastructure and Planning

The characteristics of this domain may be clear but not immediately obvious cause–effect relationships that require expert knowledge and analysis. There are a few critical example scenarios such as infrastructure planning and geo-blocking design, cybersecurity preparation and load testing, and vendor management and risk mitigation strategies, which can explain the complicated domain for the incident.

During this phase, decision-makers relied on experts to prepare the digital infrastructure. While detailed planning occurred, it failed to identify critical vulnerabilities, particularly the system's limited capacity to handle extreme load or layered cyberattacks. This reflects a misjudged complexity, where expert analysis missed key threats due to overconfidence bias and availability heuristics. Leadership should have ensured third-party audits, red-teaming, and scenario stress testing – practices aligned with sound decisions in the complicated domain.

11.4.2.2 Chaotic Domain: The Immediate Breakdown

The characteristics of this domain are not clear cause–effect relationships, urgency, and high turbulence that require rapid stabilization. There are few critical example scenarios such as sudden system failure during peak usage, misconfigured routers and high-volume DDoS activity, and decision to shut down the website to protect data integrity, which can explain the chaotic domain for the incident.

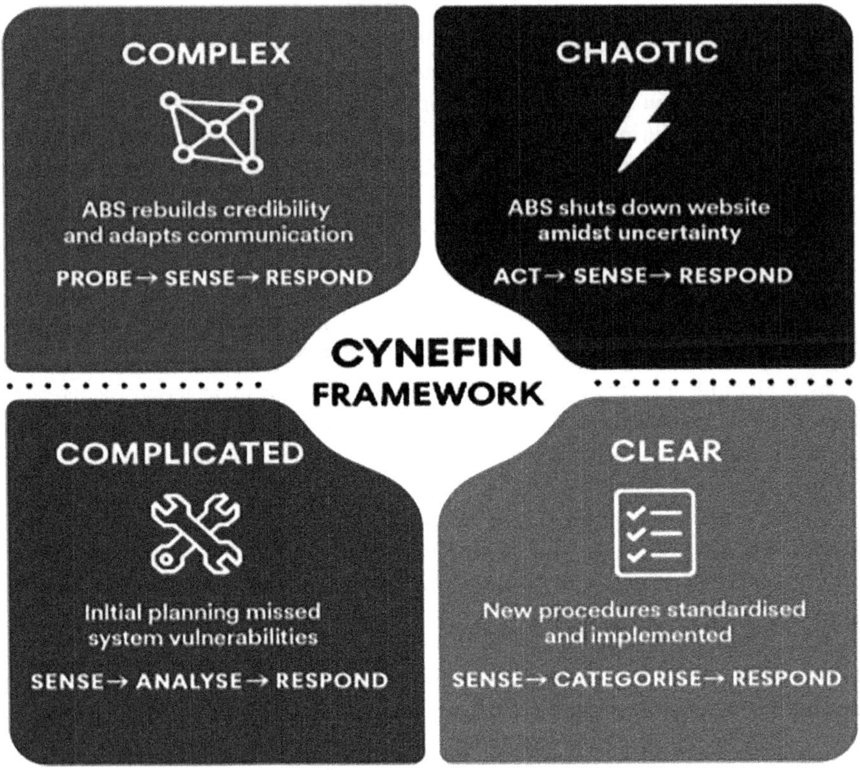

FIGURE 11.3 Crisis Management for Australian Census – A Cynefin Framework.

The situation shifted dramatically into the chaotic domain when the website crashed, and misinformation spread rapidly. In such conditions, the appropriate leadership approach is "act–sense–respond." The ABS acted swiftly by taking the system offline with an immediate intervention to protect data. However, this decision lacked an informed contingency plan. Poor communication from leadership contributed to public confusion and a loss of trust.

11.4.2.3 Complex Domain: Adaptive Strategy and Stakeholder Communication

The characteristic of this domain is to identify emergent problems with no single solution and iterative decision-making that requires experimentation and feedback. There are a few critical example scenarios such as managing public backlash and loss of trust, coordinating across government, media, and vendors, and adapting communication strategies based on stakeholder sentiment, which can explain the complex domain for the incident.

Once initial chaos subsided, the situation required complex-domain leadership: probe–sense–respond. ABS needed to test communication methods, clarify messaging, and rebuild credibility. Leaders had to engage diverse stakeholders and adapt in real time. This phase exposed a gap in adaptive leadership capacity, with the response continuing to be reactive and opaque.

11.4.2.4 Complicated (Post-Crisis Analysis) and Clear Domain (Standardization)

The characteristic of the Complicated domain for post-crisis analysis may include experts' analyse and root causes for long-term remediation, whereas characteristic of the clear domain for standardization may consider that the best practices are evident and actions can be followed standardized protocols. There are few critical example scenarios such as Alastair MacGibbon's report as post-incident reviews and implementation of enhanced cybersecurity standards and standardization of load testing and infrastructure protocols as clear domains of standardization.

As expert investigations progressed, the situation returned to the complicated domain. This led to detailed analysis of technical and procedural failures, and later, the integration of findings into the clear domain through improved operating standards and system upgrades.

11.4.2.5 Strategic Decision-Making Insights

Applying the Cynefin framework offers critical insights for strategic leadership in digital infrastructure crises. The case analysis for the Australian census outage, considering the Cynefin framework, reveals a few valuable insights and those can be described as follows:

- Contextual intelligence: Leaders must classify the domain accurately and avoid misapplying rigid decision styles.
- Timely domain shifts: Recognizing transitions (e.g., from complicated to chaotic) enables effective emergency response.
- Adaptive capability: Complex situations demand responsive, distributed leadership with continuous stakeholder engagement.
- Framework-driven learning: Shifting to the clear domain enables standardization and institutional learning from prior failure.

11.4.3 Scenario – Lion Air Crash

The Boeing 737 MAX crisis involved fatal crashes, including Lion Air Flight 610 (2018), which caused fatalities and led to a nearly two-year global grounding of the aircraft (Hradecky, 2018). The root cause was a flawed flight control system (MCAS) that relied on a single sensor and repeatedly forced the aircraft's nose down during flight. Boeing failed to inform or train pilots on MCAS and prioritized cost-cutting and market competition over safety. Regulators, particularly the FAA (Federal Aviation Administration), allowed Boeing to self-certify key systems, delaying proper oversight.

Boeing faced intense pressure to compete with Airbus' fuel-efficient A320neo and accelerated the development of the 737 MAX to retain market share. To avoid delays and additional pilot training costs, the company introduced MCAS as a software fix for aerodynamic changes. Despite internal warnings about its safety, executives prioritized rapid rollout and sales over thorough validation, creating a decision-making environment driven by market urgency rather than safety concerns.

The Lion Air Flight 610 crash provided Boeing with undeniable links between the MCAS system and the tragedy, but they still opted against grounding the 737 MAX aircraft (Ghaisani ET AL., 2024). Boeing provided pilots with basic instructions while giving commercial interests precedence over safety measures. Boeing's leadership ignored necessary actions due to cognitive and organizational biases, which resulted in another deadly accident while highlighting significant shortcomings in their crisis management and risk communication.

This use case analysis applies the Cynefin framework and illustrates how poor judgement, cognitive biases, and weak regulatory checks can result in catastrophic outcomes in high-stakes industries. The Cynefin framework is selected to guide Boeing's leadership in navigating the crisis. This framework classifies decision-making contexts into five domains – Clear, Complicated, Complex, Chaotic, and Disordered that could enable leaders to tailor strategies to situational complexity.

The Cynefin framework is ideal for the selected use case scenario because it accounts for uncertainty and complexity, unlike traditional risk management models that assume predictable outcomes, counters cognitive biases (e.g., overconfidence) by promoting iterative probing and diverse stakeholder input, aligns with Boeing's innovation value by encouraging experimental solutions, such as simulator-based MCAS testing, and supports strategic leadership by providing a structured yet flexible approach to crisis management.

The Lion Air crash scenario falls within the Complex domain, characterized by:

- High uncertainty due to incomplete crash data and evolving evidence
- Conflicting stakeholder interests (safety vs. commercial priorities)
- Delayed feedback loops from ongoing investigations
- No clear "best practice" response, as MCAS issues were new

In Complex domains, cause-and-effect relationships are only discernible in hindsight. The Cynefin framework advocates a "probe–sense–respond" approach, encouraging experimentation, data collection, and adaptive responses. This contrasts with Boeing's initial linear approach, which underestimated the problem's severity and relied on outdated assumptions. The Probe–Sense–Respond approach of the Complex Domain in the Cynefin framework can be explained as follows:

1. Probe: Perform independent audits of MCAS using pilot simulator tests to replicate crashes and involve third-party experts to evaluate reliability and identify risks.
2. Sense: Analyse crash data, pilot input, and stakeholder views to identify systemic issues. Employ cross-functional teams to question assumptions and reduce confirmation bias.
3. Respond: Complement adaptive actions like grounding the 737 MAX, updating software to remove single-sensor reliance, and improving MCAS pilot training.

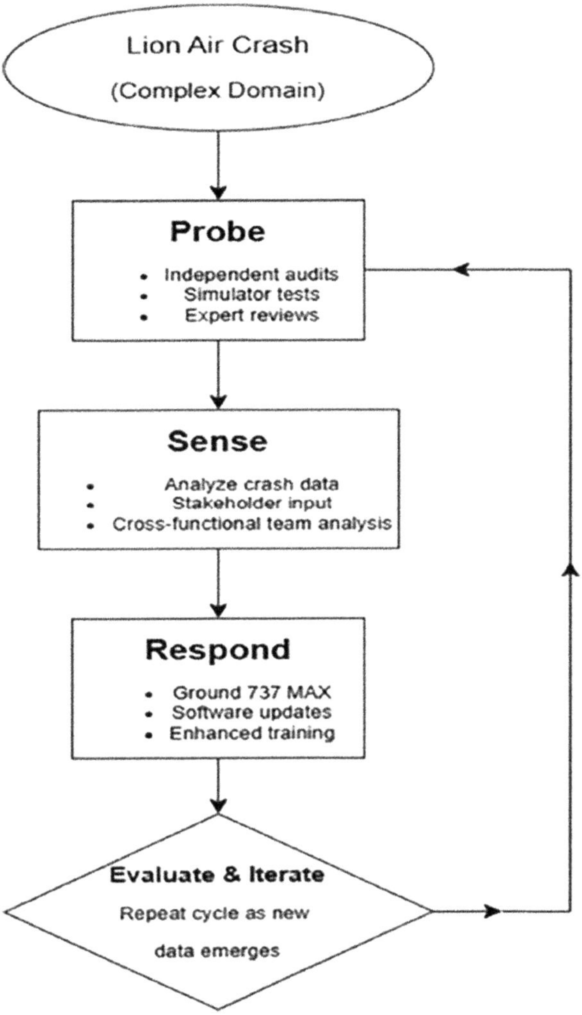

FIGURE 11.4 Probe–Sense–Respond Approach of Complex Domain in Cynefin Framework.

This iterative cycle, as shown in Figure 11.4, ensures decisions evolve with new evidence, aligning with safety and integrity values. A conceptual diagram of this structure is presented below.

Table 11.1 shows the application of the Cynefin framework to Boeing's Post-Lion Air crash decision context. The table defines three phases such as probe, sense, and respond for the Boeing context with suggested actions that clearly addresses the situations.

The Cynefin framework demonstrates that Boeing's linear method was inadequate for managing the complex situation of the 737 MAX crisis. Leadership failed to adjust to the situation and instead simplified the MCAS problem while neglecting the

TABLE 11.1
Application of Cynefin Framework for Lion Air Crash Scenario

Cynefin Phase	Boeing Context	Suggested Action
PROBE	Lack of internal clarity and risk data on MCAS	Run simulations and gather confidential pilot evaluations while internal communications
SENSE	Conflicting interpretations of MCAS safety	Using FDR data trends along with incident history and pilot report to identify persistent malfunctions
RESPOND	Missed opportunity for decisive leadership	Halt the entire aircraft fleet for a limited duration while maintaining open communications lines with stakeholders

TABLE 11.2
Comparison of Boeing's Response and Cynefin-Guided Action

Aspect	Boeing's Response	Cynefin-Guided Action
Problem Framing	Linear, reactive	Emergent, exploratory
Stakeholder Input	Narrow, internal	Broad, multi-level engagement
Decision-Making Style	Rigid, delay-oriented	Adaptive, feedback-driven
Outcome	Second crash, loss of public trust	Likely risk mitigation and enhanced stakeholder trust

broader systemic dangers. Implementing a "probe–sense–respond" strategy would have driven deeper investigation and testing procedures while fostering stakeholder partnerships and evidence-based responses such as fleet grounding and transparent information sharing, which would have established responsible long-term risk management practices. The comparison of Boeing's actual post-crash leadership response and a Cynefin-guided strategic actions can be described as shown in Table 11.2.

Cynefin mitigates leadership biases like overconfidence, groupthink, and resistance to change by highlighting ambiguity and promoting exploration, scenario testing, and diverse viewpoints, key elements in effective crisis leadership. Once Cynefin clarifies the situation and intervention needs, Boeing's leadership could use structured decision tools like AHP/TOPSIS to effectively compare and select implementation options.

11.5 CONCLUSION

This chapter forms a holistic approach to explain leadership and strategic decision-making following the Cynefin framework that supports resilient, evidence-based decision-making across all stages of crisis response. The insights gained are not only applicable to national infrastructure projects like the Chernobyl nuclear operation, the Australian Census, and Lion Air crash, but also extend to broader strategic contexts where high uncertainty, public scrutiny, and technological risk converge. By embedding this decision framework into organizational practice, organizations like Chernobyl or the Australian Census Website can enhance their foresight, improve their response to emerging threats, and rebuild public trust in digital service delivery.

The Cynefin framework provides insights into the complex decision-making process associated with two use case scenarios. This model categorizes situations into five domains – simple, complex, complicated, chaotic, and disordered that support stakeholders to differentiate cause-and-effect relationships and respond appropriately. By mapping scenarios to these domains, decision makers can adopt tailored approaches

The Cynefin framework serves as a powerful lens to assess Chernobyl, Australian Census Website, and Lion Air leadership considering the incident. It highlights how misjudging the nature of the situation coupled with poor communication and inflexible responses that intensified the crisis. Domain-aware leadership, grounded in contextual sense-making, could have enabled more effective planning, incident response, and public engagement. Going forward, embedding the Cynefin approach into strategic decision-making processes can strengthen resilience in digital infrastructure and improve leadership under uncertainty.

BIBLIOGRAPHY

Australian Bureau of Statistics (ABS). (2017). *Review of the events surrounding the 2016 Census online system failure*. Canberra: Australian Bureau of Statistics. www.abs.gov.au/websitedbs/D3310114.nsf/home/2016+census+online+system+review.

Ayene, M. Boeing 737 max 8 operations suspended in Australia after Ethiopian airlines crash. www.abc.net.au/news/2019-03-12/boeing-737-max-8-suspendedoperations-australia/10894426.

Bennett, B., Repacholi, M., Carr, C., Bennett, B., Repacholi, M., & Carreditors, C. (2006). *Report of the UN chernobyl forum expert group "health", health effects of the Chernobyl accident and special health care programmes*. World Health Organization. English, Book, 92-4-159417-9 9789241594172.

Beresford, N. A., Fesenko, S., Konoplev, A., Skuterud, L., Smith, J. T., & Voigt, G. (2016). Thirty years after the Chernobyl accident: What lessons have we learnt?. *Journal of Environmental Radioactivity*, 157, 77–89. https://doi.org/10.1016/j.jenvrad.2016.02.003.

Chernousenko, V. M. (2012). *Chernobyl: Insight from the inside*. Springer Science & Business Media.

French, S. (2013). Cynefin, statistics and decision analysis. *Journal of the Operational Research Society*, 64(4), 547–561, doi: 10.1057/jors.2012.23.

Ghaisani, A., Sefriani, S., & Safitri, A. A. (2024). The lion Air JT610 crash due to lack of pilot training: Is there responsibility for Indonesia?. *Uti Possidetis: Journal of International Law*, 5(2), 303–332.

Harper, L. L., Desy, J., Davis, M., Weeks, S., & McLaughlin, K. (2024). Personal career decisions during medical training are not complicated, they are complex. *Medical Education*, 58(9), 1042–1048, doi: 10.1111/medu.15466.

Hradecky, S. (2018). *Crash: Lion B38M near Jakarta on Oct 29th 2018, aircraft lost height and crashed into Java Sea, wrong AoA data*. The Aviation Herald.

International Nuclear Safety Advisory Group. (1992). *The chernobyl accident: Updating of INSAG-1: INSAG-7: A report (No. 7)*. International Atomic Energy Agency.

Joint Committee of Public Accounts and Audit (JCPAA). (2017). *Cyber security of the 2016 Census*. Parliament of Australia. https://www.aph.gov.au/Parliamentary_Business/Committees/Joint/Public_Accounts_and_Audit/Census2016.

Krukovskyi, P., Metel, M., & Polubinskyi, A. N. A. T. O. L. I. I. (2018). New save confinement as a barrier between destroyed reactor of the chernobyl NPP and environment (Design, Purpose, Modeling). *WSEAS Transactions on Environment and Development*, 14, 197–204.

Linsky, M., & Heifetz, R. A. (2002). *Leadership on the line*. Harvard Business School.

Medvedev, G., & Sakharov, A. (1991). *The truth about Chernobyl*. Basicbooks, A Division of Harpercollins Publishers.

Nachbagauer, A. (2021). Managing complexity in projects: Extending the Cynefin framework. *Project Leadership and Society*, 2, 100017.

Snowden, D. J., & Boone, M. E. (2007). A leader's framework for decision making. *Harvard Business Review*, 85(11), 68–76. https://hbr.org/2007/11/a-leaders-framework-for-decision-making.

Topham, G. (2019). Ethiopian flight 302: Second new boeing 737 to crash in four months. *The Guardian*, 10.

United Nations Scientific Committee on the Effects of Atomic Radiation. (1988). *Report of the United Nations scientific committee on the effects of atomic radiation (No. INIS-mf--13182)*. United Nations Scientific Committee on the Effects of Atomic Radiation.

van Beurden, E. V., Kia, A., Zask, A., Dietrich, U., & Rose, L. (2013). Making sense in a complex landscape: How the Cynefin framework from complex adaptive systems theory can inform health promotion practice. *Health Promotion International*, 28, 73–83, doi: 10.1093/heapro/dar089.

World Nuclear Association. (2022). RBMK reactors – Appendix to nuclear power reactors – world nuclear association. https://world-nuclear.org/information-library/appendices/rbmk-reactors#post-accident-changes-to-the-rbmknbsp.

12 Innovative Decision-Making for Human and Machine Intelligence

12.1 INTRODUCTION

In today's world, relying on past experiences for innovation may not be enough. Lean methodology recommends confirming ideas with real-time data analysis when making rational decisions. Data visualizations make it simpler for processing critical information, supporting diverse decision-making. Established organizations may avoid innovative ideas that can threaten their core operations. Seeking an external perspective following innovative decision-making (IDM) methods can support them in exploring new ideas more comprehensively and mitigate potential risks. Professional teams can utilize agile and IDM frameworks to expedite the decision-making process, even in complex and uncertain situations. Professionals must rationalize their decisions through integrative consultation, evidence-based options, and extracted innovative information from advanced mechanisms.

In a fast-changing world, decision-making is becoming more complex than ever. Understanding the context and socio-technical effects is important to make informed decisions, particularly for human–machine interaction. This includes a few stages: contextual analysis, innovative solutions through data analytics, stakeholder analysis, and judgemental decision-making that encourage logical evidence and continuous monitoring. Stakeholder analysis identifies key stakeholders and their potential involvement in making decisions. Further, professionals can engage data analytical tools to explore innovations following established frameworks with diverse perspectives that can avoid cognitive biases. They can apply a divergent approach to brainstorm with critical thinking that enables them to analyse available resources and encourage innovative thinking. After generating ideation, they can follow a convergent approach that supports them narrow down topics and explore new ways to solve critical problems. Advanced technologies such as artificial intelligence (AI), big data analytics, and cloud computing can enhance the decision-making process through evidence-based solutions. Finally, professionals can assess alternatives with logical matrices based on efficiency, feasibility, maintainability, social impact, and organizations' goals.

Professionals can execute analytical processes following IDM methods, exploring agile and innovative solutions by exploring opportunities and conducting experimentation. Innovative decisions need to accommodate stakeholder engagement, contextual

DOI: 10.1201/9781003685449-12

data analytics, and the dedication of experts in complex situations. They can allow adaptation to encourage agility through innovative solutions and continuous learning. The IDM is based on reviewing past experience and contextual information to extract valuable insights and new solutions. To address uncertainties, leaders must be proactive to make innovative decisions considering diverse perspectives and incorporate innovative problem-solving mechanisms.

In uncertain situations, professionals can carefully balance organizational demands, environmental considerations, technical constraints, ethical responsibilities, and available resources, which are essential aspects of IDM, particularly in fast-paced environments where rapid responses are expected. They can also integrate rational analysis with ethical reasoning to justify and execute informed decisions effectively. To support this process, leaders can follow established IDM frameworks that provide systematic, sequential guidance in critical scenarios. By leveraging insights generated from these frameworks, decision-makers gain actionable recommendations and can select the most suitable option from available alternatives. In today's fast-moving world, organizations must harmonize the urgency of action with thorough, innovative analysis, enabling swift, well-informed decisions that account for limitations, opportunities, and competitive advantage.

Agile and IDM embraces collaboration, credibility, transparency, iteration, adaptability, value co-creation, and noble decisions in professional activities. Stakeholders can contribute to the decision-making process through close collaboration, innovations through iterations, data analytics, and evolving solutions. Moving ahead needs agility, innovations, customer-centric, and avoiding relying on past success. Agile is a mindset, not just a process, which supports collaboration and prevents command-and-control to make an equitable environment in the decision-making process. The IDM can maintain professionals' autonomy and equally equip them through advanced analytical techniques with proper justification. They can follow principles on the critical thinking process that identify constraints and biases to overcome critical situations.

Innovation management converts ideas into valuable assets, strengthens customer satisfaction, and fuels business growth, often leading to new products and even the transformation of entire industries. Its success relies on sustained and skilful investment, as innovation follows a structured journey of discovery and commitment. This journey begins with knowledge, where awareness of new possibilities is sparked, followed by persuasion, as insights shape perceptions and attitudes. The innovative decision stage involves choosing whether to adopt or reject the innovation, leading to implementation, where ideas are translated into actions. Finally, confirmation entails reflecting on outcomes to affirm or refine the chosen path. As the principle suggests, effective innovation requires breaking problems down into fundamental truths and building solutions upward from there.

IDM is intrinsically uncertain in unexplored domains, where outcomes are fluid and success must be continually reinterpreted as reliable data are limited or unavailable. In such contexts, plans function more as navigational guides than fixed routes, offering direction without guaranteeing precision. Leaders must therefore move beyond reliance on data-intensive decision-making and develop the capacity to operate effectively amid ambiguity, recognizing that a nuanced understanding of the innovation process itself is vital. Acting decisively and prioritizing timely action over hesitation becomes critical for sustaining competitive advantage. Organizations that clearly choose between quality and speed, deliberately pursue either incremental or transformative innovation, and balance decisive leadership with team empowerment are better positioned to outperform their competitors.

IDM necessarily involves pushing boundaries and engaging with uncertainty. A clear and well-defined understanding of the problem must precede solution design, as prematurely committing to an initial plan can result in wasted time, effort, and resources, often compounded by the sunk cost fallacy. Effective innovation therefore depends on the ability to make rational decisions at speed. In large organizations, this process can be strengthened by seeking cross-functional input and obtaining early approvals, which accelerates execution, reduces resistance to change, and distributes accountability for unfavourable outcomes. Given the competing pressures faced by decision-makers, even routine choices can become protracted; in such circumstances, a timely but imperfect decision frequently delivers greater value than a perfectly reasoned decision made too late.

12.2 HUMAN AND MACHINE INTELLIGENCE

Human intelligence is a cognitive process in fulfilling humans' expectations or requirements from social or machine interactions. It develops a mode for people when they perceive experiences through interactions. Human–computer interaction or social influence can develop their cognitive and emotional engagements through a sense of presence, efficacy, subjective norms, attitude, satisfaction, and trust, consequently impacting behavioural responses. Peoples' mental state confirms human intelligence that can commit to perform certain actions after fulfilling specific expectations from the environment.

Organizations are adopting AI and big data analytics in decision-making processes that involve sensory information and function as invisible actors to influence their cognitive and emotional involvement through machine intelligence. AI co-creates values to captivate their decisions through personalized recommendations, a sense of immersion, and mental imagery. AI engages collaborative processes, where professionals can get valuable insights through advanced data analytics platforms through sensory information in human–computer interaction, enhancing intelligent and interaction capabilities. In another context, big data analytics can support organizations to extend analytical processes through advanced data analytical tools.

AI and big data allow organizations to extend predictive analysis that explores user requirements through machine intelligence using predictive and explainable models. Furthermore, predictive and responsive analytics can make it possible to understand organizations' goals, peoples' preferences, and contextual inclinations to recommend alternatives for decision-making. These AI-generated personalized recommendations and valuable insights through big data analytics can engage in a new way to encourage professionals through rational decision-making by extracting valuable perceptions and enhancing human decisions through machine intelligence.

12.3 INNOVATION DECISION-MAKING PROCESS

IDM evaluates creativity in four domains and devises a customized strategy to improve decision-making. The first step of decision-making is identifying the problem, which may require information gathering and problem diagnosis. In the exploration phase, brainstorm alternative solutions for the decision situation. In the search phase, values are assigned to each alternative after gathering and processing information and weighing the pros and cons. Selection is committing to an alternative and implementing it. Good decisions are often delayed or not made due to indecisiveness or ignoring practicalities.

Despite investing in digital tools, agile methodologies, and lean strategies, businesses struggle with inadequate processes and models for quick and effective innovation decisions. Discovery-driven innovation requires efficient decision-making to identify risks and avoid initiative failure. Diverse teams excel at identifying opportunities and risks, but experimental organizations need four key perspectives in decision-making. IDM assesses creativity in four areas as shown in Figure 12.1 and offers a personalized action plan to enhance decision-making skills for more effective decisions.

Figure 12.1 illustrates the process flow for IDM, where professionals can brainstorm to find available resources to fix objectives, search appropriate content using advanced technologies, identify problems, assess multiple options, and finally select the best option in making decisions.

IDM encompasses value delivery, empowerment, transparency, iteration, cooperation, and flexibility. Every stakeholder participates in the decision-making procedure; decision-making authority is shared, teams operate continuously, and strategies

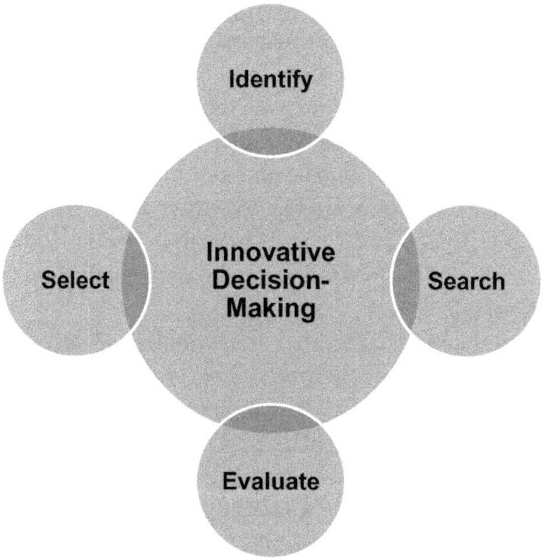

FIGURE 12.1 IDM Process.

change as conditions alter. To stay ahead, one must be agile, take measured risks, put the needs of the client first, and not rely too much on past achievements.

The IDM can apply big data analytics to analyse large volumes of data following the organisation's requirements (Shahid & Sheikh, 2021). Big data analytics is collecting, coordinating, and analysing huge volumes of data from various sources to hunt for patterns, relationships, and information that drive actions. Big data analytics integrates technologies such as machine learning, data mining, real-time processing, and predictive analytics with the aim of facilitating high-velocity and smart decision-making in complex environments.

Big data involves the four Vs, such as Volume, Velocity, Variety, and Veracity, which make decision-making capable of dealing with various data streams like sensor data, text reports, weather forecasts, and social signals. Timeliness comes from the ability in the model to keep updating analysis with data that is newly arriving, facilitating iterative decision loops.

12.4 IDM MODEL

12.4.1 Roger's Model

The innovation-decision model can be seen as a journey through an evolving landscape in which new possibilities are gradually discovered and embraced. It progresses through a series of stages: Awareness, where the presence of an innovation is first recognized: Assessment, in which attitudes and perceptions towards the innovation are formed; Choice, where a decision is made to accept or reject it; Adoption, involving the actual implementation of the innovation; and Affirmation, where ongoing feedback and evidence are sought to confirm that the chosen course is appropriate.

In Rogers' model, professionals follow a process flow as shown in Figure 12.2, executing IDM for complex situations (Rogers, 2003). They initially reflect on existing practices, identify problems, recognize levels of innovativeness, and consider prevailing social norms. In the next stage, they acquire knowledge and situational awareness that enable them to conduct immediate tasks using innovative ideas. This

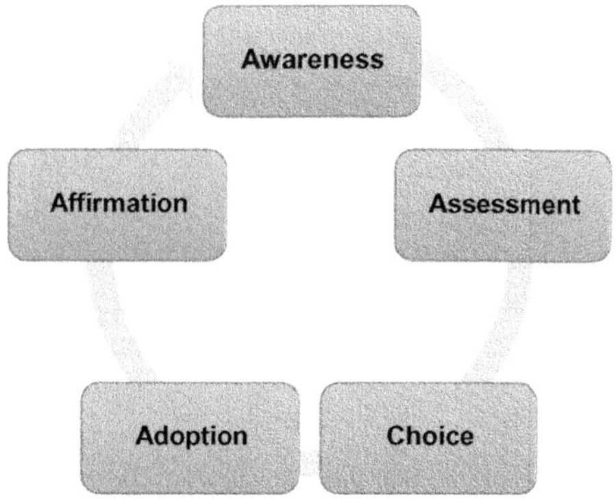

FIGURE 12.2 Process Flow for Roger's Model.

is followed by a comprehensive evaluation in which the perceived characteristics of the innovation are analysed, and choices are shaped by factors such as compatibility, observability, and reliability. Subsequently, professionals decide whether to adopt or reject the available options. Finally, they seek confirmation by affirming their decision to adopt innovative solutions or reconsidering their choice.

The IDM approach delivers several important advantages. By proactively testing and simulating decisions with available data, it helps reduce decision paralysis and encourages timely action rather than prolonged analysis. The approach is inherently adaptive, acknowledging that decisions can be refined as the latest information emerges. It also supports effective risk management by discouraging rushed or poorly informed choices, ensuring that actions are not taken without adequate evidence. Moreover, ongoing data collection strengthens organizational learning, promoting a culture of adaptability, reflection, and increasingly informed decision-making.

Relying on a single expert's final decision can suppress open discussion and introduce bias, discouraging others from voicing alternative perspectives. To avoid superficial solutions, advocates of different options can broaden their viewpoints and remain open to challenges, as a willingness to resist easy compromise can often spark more innovative outcomes. Agile leaders recognize when further learning and experimentation are required, and new professional teams must substantiate their ideas with evidence of superior value to end-users. Diversity of participation alone is insufficient; organizations need a competitive marketplace of ideas where assumptions are evaluated, risks are surfaced, and the most promising initiatives can be identified and pursued.

This approach emphasizes data-driven patience, drawing on the steady flow of evidence while allowing time for thoughtful judgement in decision-making. It integrates consensus-building with real-time insights, enabling rational decisions to adapt continuously as the latest information emerges. By balancing immediate pressures with long-term considerations, it seeks harmony between short-term action and future impact. Risk is mitigated through patience combined with informed choices, while lean principles encourage simplicity as a pathway to greater value and sustainable growth. Agile decision-making supports navigation through complexity by acknowledging both what is known and unknown, and a continuous quest for knowledge supported by web-based intelligence that guides strategic thinking amid uncertainty.

Optimal decisions are achieved by simulating real-time scenarios, systematically gathering and analysing data, and exercising patience throughout a dynamic and IDM process. Professionals rely on data and simulations to guide their choices and meet deadlines, striking a balance between timely action and informed judgement. This approach reduces decision paralysis by keeping decision-makers actively engaged, enabling them to act without becoming stalled by excessive analysis. It also promotes flexibility, recognizing that decisions can evolve as the latest information emerges. By discouraging hasty, poorly informed choices, the approach supports effective risk mitigation, while continuous data collection and reflection foster a culture of learning and adaptability that strengthens future decision-making across the organization.

IDM is a workshop-based methodology designed to help teams improve decision-making through the use of innovative tools and techniques. Leaders frequently confront complex choices shaped by incomplete information, cognitive biases, and far-reaching consequences, where sound decisions align with organizational objectives and ethical standards, while poor decisions can be costly. Decision-making often places organizations at a crossroads, balancing the chosen path against the uncertainty of the road not taken. Addressing this challenge, an innovative approach known as an IDM integrates real-time actions with ongoing information gathering, enabling decisions to progress without being constrained by regret or speculation. An effective IDM is central to leadership and management, as it involves selecting among alternatives to guide organizational direction and drive sustained progress.

12.4.2 The Observe–Orient–Decide–Act Loop Framework

The Observe–Orient–Decide–Act (OODA) loop, developed by U.S. Air Force strategist John Boyd, is a rapid decision-making cycle designed for fast-paced and uncertain environments (Boyd, 1996). It is especially applicable in engineering settings where evolving risks, time pressure, and limited information must be addressed through iterative and responsive action.

The OODA Loop can be used to illustrate how IDM may be applied within organizations, particularly in critical situations. In the case of the Hyatt Regency project, construction was well underway by February 1976, and no incidents were reported for two years. However, in October 1979, a major structural failure occurred when more than 2,700 square feet of the hotel's glass atrium roof collapsed during construction (Pfrang & Marshall, 1982). Although this incident did not result in injuries, it should have served as a clear warning of deeper systemic weaknesses. Instead, the absence of a formal investigation or comprehensive review of the design and construction processes meant that management and engineering teams missed a crucial opportunity to identify underlying issues, challenge design assumptions, and strengthen communication protocols.

This case is a good example of the lack of responsive decision-making tools in a changing building process. Linear models do not know how to manage early signs of risk in environments characterized by changing information landscapes, low information, and time pressure. Instead, agile approaches that focus on observation, continuous feedback, and quick response (such as the OODA Loop) can make a difference. The OODA phases for the Hyatt use case can be described as shown in Table 12.1.

If the Hyatt Regency team had adopted an OODA-based approach following the 1979 atrium roof failure, a radically different trajectory may have emerged. Here is how the loop could have functioned in practice:

I. Observe: The roof collapse offered a clear, observable deviation from expected performance. Material anomalies and installation discrepancies should have triggered a data-gathering phase – recording construction flaws, engaging structural teams, and reviewing shop drawings.
II. Orient: With observed data in hand, engineers and project managers could have contextualized the failure – not as an isolated event but as an indicator of broader structural concerns. This orientation phase would involve

TABLE 12.1
OODA Phases of Hyatt Use Case

Phase	Description	Application of the Hyatt Case
Observe	Gather situational data and warning signals	1979 roof collapse, material performance issues, fabrication delays
Orient	Analyse context, assumptions, and stakeholder positions	Reassess safety protocols, communication gaps, and prior engineering assumptions
Decide	Choose an adaptive strategy based on current insight	Implement formal design review, independent audit, and risk escalation protocol
Act	Execute strategy, monitor outcome, feed into the next loop	Conduct recalculation, document decisions, establish peer-review mechanisms

consulting cross-functional teams, reflecting on construction pressures, and challenging assumptions about previous design decisions (including the then-unchallenged rod system).

III. Decide: The appropriate adaptive strategy would have included mandating recalculations of key suspended elements, enforcing a project-wide review of all hanging structures, and requiring written documentation for design modifications. Teams may also have considered halting parts of the project until verification processes were complete.

IV. Act: Investing in design would have required peer reviews, third-party audits, directing voices, providing book of recordable design; none of which occurred in any formal or traceable manner throughout the project. That is what the verbs would do in the next round of the loop: they would feed into it, as a self-correcting process of governance.

The OODA Loop is cyclical and iterative, meaning new data continuously informs orientation and decision stages. This contrasts with traditional waterfall models where decisions are fixed early in the process and rarely revised. The visual framework of OODA Loop in the Hyatt use case can be illustrated as shown in Figure 12.3.

The OODA Loop would have changed the game in the Hyatt initiatives by injecting real-time responsiveness, proactive risk assessment, and fast feedback loops. Unlike static frameworks such as the Rational Planning Model, OODA enables adjustments to be made dynamically – vital in fast-paced, quickly shifting environments, such as construction projects at a large scale. This model is also consistent with multi-stakeholder communication because each loop promotes the integration of different know-hows, before triggering actions.

The framework was chosen for its adaptive, rapid-response capability, which directly addresses the missed opportunities following the 1979 roof collapse. Unlike traditional decision models that assume linear progression, the OODA Loop is designed to operate effectively under uncertainty, enabling early detection of emerging risks and the reassessment of flawed mental models. While alternative approaches such as

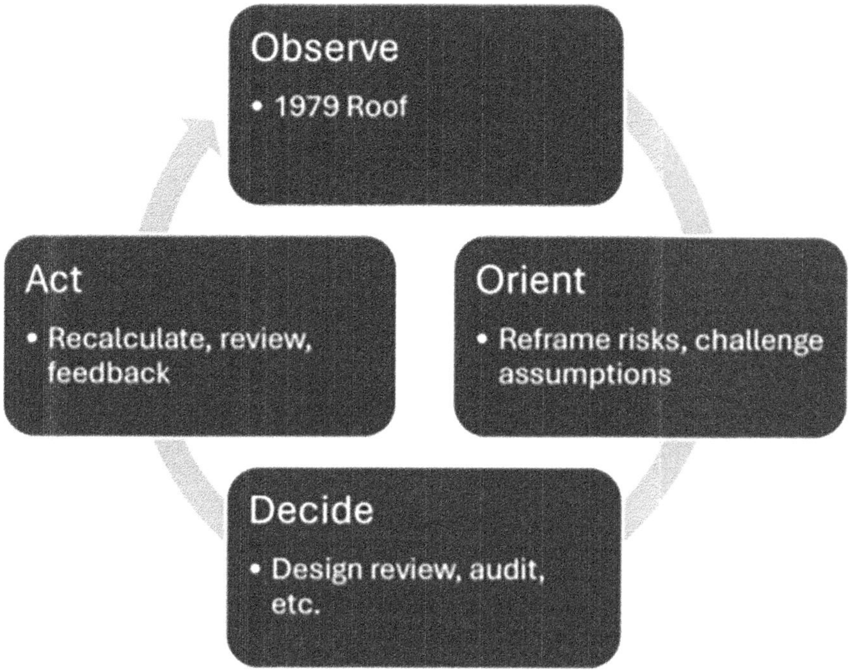

FIGURE 12.3 Visual Framework for OODA Loop in Hyatt Use Case.

SCRUM-based engineering reviews or Design Thinking also provide agility, they are more appropriate for product development or innovation-focused contexts. In contrast, OODA is particularly well suited to time-critical, safety-sensitive environments such as construction, where minor oversights can quickly escalate into major failures. Its origins in military strategy further reinforce its suitability for crisis management, a critical consideration in the Hyatt case, where early warning signs of structural failure were overlooked.

Each of the selected decision-making models (IDMs) such as the Delphi Method, the Cynefin Framework, and the OODA Loop, can deliberately be chosen to align with distinct failures evident in the Hyatt Regency walkway collapse, including breakdowns in group consensus, strategic misclassification of the situation, and insufficient responsiveness to early warning signals. The rationale for this selection lies in the unique strengths each model offers, both theoretically and in relation to the contextual complexities of the case, as well as their ability to address the cognitive biases and process-related deficiencies that contributed to the incident.

12.4.3 Double-Loop Learning Method

Double-Loop Learning is an IDM method introduced by Chris Argyris and Donald Schön (Argyris & Schön, 1978). It describes how professionals and organizations can learn from experience, not just by rectifying errors, but by probing mechanisms to

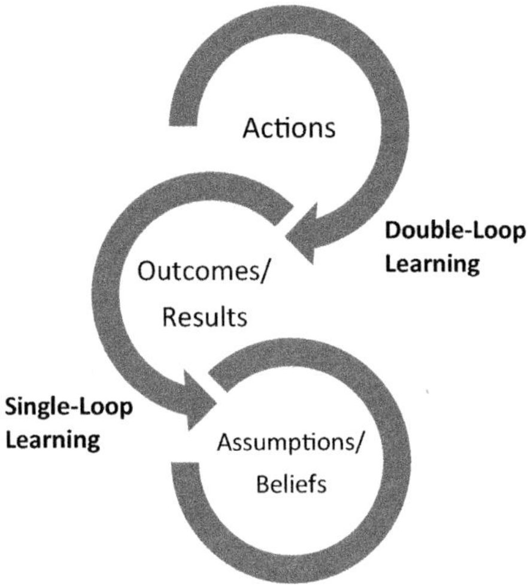

FIGURE 12.4 Double-Loop Learning Method.

review in detail assumptions that can create those errors in uncertainty. This process yields it exceptionally effective for IDM.

The Double-Loop Learning operates through two interconnected feedback loops as shown in Figure 12.4. The first loop focuses on action, outcome, and correction, identifying errors and adjusting actions to better achieve existing objectives. The second loop goes deeper by linking outcomes to underlying assumptions and redesign, questioning the governing variables that shape decisions such as goals, policies, norms, risk tolerance, and success criteria (Argyris, 2002). It is within this second loop that genuine innovation emerges, as organizations challenge and redefine the fundamental frameworks guiding their actions.

This chapter applies the Double-Loop Learning method to extend decision-making analysis for a critical use case scenario. At the beginning of March 28, 1979, a small failure in the secondary cooling system of the Three Mile Island (TMI) nuclear power plant in Pennsylvania initiated a series of technological breakdowns and operator disorientation (Ahearne, 1986). An uncontrolled, stuck-open pressure-relief valve that the control room operators did not know about, and their responses, under the guidance of faulty instrument readings, unintentionally exacerbated the coolant loss and reactor core overheating. The compounding pressures resulted in a partial meltdown of the plant's Unit 2 reactor, the most severe accident in U.S. commercial nuclear power history. Though the first radioactive release was minor, having no detectable effect on public health, the accident generated alarm in the industry and society, destroying confidence in nuclear safety and initiating widespread changes in reactor operation, training, and regulation.

TMI has become a milestone in decision science, organizational behaviour, and cognitive psychology. Accident investigations found mechanical and design flaws and exposed failures in human cognition, communication, and group performance. Misperception of ambiguous instrument readings by the control room personnel, after flawed mental models notwithstanding mistakes when stressed, and other cognitive fallacies, demonstrated how human factors compound technical failure into a full-fledged crisis. Likewise, leadership and crisis communication failures at TMI eroded public confidence, which helped to show organizational setting and decision-making in high-risk contexts. In effect, the TMI meltdown vividly demonstrated that catastrophes in tightly coupled, complex systems seldom are strictly "technical" – they are related to how people perceive, make decisions, and act under pressure.

In the case of organizational reaction to the TMI accident and its effects, the Double-Loop Learning framework is employed in this stage. Single-Loop Learning, where mistakes are found and addressed based on existing norms, differs from Double-Loop Learning because the latter challenges implicit assumptions, values, and policies. Double-Loop Learning enables organizations to reflect not only on actions but also on the belief systems shaping the actions, resulting in adaptive and transformative change.

Where high-risk, tightly coupled systems already in place, such as nuclear power stations, are involved, the traditional learning loop with technical rectification alone can be found wanting. TMI is an example of a deeper failure not only in operation but in how the organization perceived problems, responded under pressure, and created learning. Hence, a double-loop model is more appropriate to deal with cognitive, cultural, and structural flaws that were the root of the crisis.

Following the TMI accident, a wide-ranging package of reforms came into existence. The regulations were tightened, training of operators was extended, and reactor design was reviewed. But the question still has to be asked: Did the response amount to a paradigmatic shift in the way that the nuclear industry thinks and acts, or was it a matter of single-loop correction?

Based on a double-loop framework as shown in Figure 12.4, the first operator reaction during the accident had its roots in flawed mental models. Operators misinterpreted signs, over-relied on procedures, and made assumptions about the state of the system that were not challenged until it was too late. These are archetypal symptoms of a requirement for Double-Loop Learning in which assumptions such as "valve commands always work" or "emergency cooling is bad unless proven otherwise" need to be brought to the surface and examined.

Post-crisis, the utilities and the U.S. Nuclear Regulatory Commission (NRC) attained improvement in instrumentation, technical controls, and simulation-based training (NRC Fact Sheet, 2018). While these steps are a fine example of Single-Loop Learning, evidence of true Double-Loop Learning, such as re-evaluation of hierarchical decision-making structure, communication channels, and over-reliance on automation, was less.

There are, however, some areas that reflect double-loop thinking. For example, the Institute of Nuclear Power Operations (INPO), created in the aftermath of the accident, emphasized a culture of safety that supported critical thinking, openness, and peer review. This move towards a questioning approach, where frontline personnel are enabled to challenge assumptions, is consistent with the philosophy of Double-Loop Learning. Besides, scenario-based training and simulation began to simulate uncertain, ambiguous events, challenging trainees to move away from routine procedures towards reflective judgement.

Nonetheless, Double-Loop Learning implementation was not consistent. Cultural resistance, regulative constraints, and institutional refusal to confront larger faults too often suppress more drastic changes. Reactions most often depended on technical fixes, new meters, and better alarms instead of a cognitive or structural shift.

The double-loop model was selected because it can be applied to crisis-prone, complex systems like nuclear power. Technical malfunctions have happened at TMI, but the underlying causes were based on human thought, organizational culture, and systemic assumptions. A model that does no more than correct procedural mistakes will not fully explain or lead to the next failure being prevented.

Double-Loop Learning allows an organization to question its governing variables. Double-Loop Learning goes beyond "What did we do wrong?" to inquire, "Why did we think that was the right thing to do?" Double-Loop Learning is therefore especially well-suited to the TMI situation, in which operators' actions were guided by flawed beliefs embedded within training, norms, and system design.

The model also fits with the broader objective of learning from crises. It does not assume that errors are isolated aberrations but sees them as opportunities for radical change. In an industry where errors can lead to catastrophic outcomes, only deep, reflective learning can ensure long-term resilience.

Amongst other agile models like learning organizations or adaptive leadership, Double-Loop Learning is the most conscious model for surfacing and modifying implicit mental models. While all three promote flexibility, Double-Loop Learning specifically addresses surfacing and challenging in-built assumptions and thus is best applied to root-cause organizational reflection.

Double-Loop Learning draws on organizational learning theory and behaviour science and draws heavily on the decision-making process. It is a development of the distinction between espoused theory (what people say they believe) and theory-in-use (what they do) and identifies the gap between formal rules and real behavioural actions. In TMI, this kind of gap manifested itself in the usage of procedures by the operators, which were inconsistent with the reality of the system.

The model also overlaps with cognitive psychology, namely in the identification of mental models' contribution to determining action and perception. TMI operators operated on an incorrect mental model of the reactor system, believing that the Pilot-Operated Relief Valve (PORV) had closed and that belief constrained their options. Not questioning that assumption is characteristic of a failure, not of knowledge but of reflective inquiry.

Taking the bigger picture, Double-Loop Learning requires organizations to make reflection a part of their culture. This is not simply about post-mortems but about habituative practices that increase dissent, evaluate orthodoxy, and incentivize curiosity. In High-Reliability Organizations (HROs), these are core to resilience. The TMI case demonstrates to us what happens when there is no reflection, when assumptions are never evaluated, and signals are ignored.

Finally, Double-Loop Learning offers a rich context for examining not just the short-term failures but also the subsequent reforms following the TMI accident. Double-Loop Learning compels the conversation away from procedures and into the realm of values, culture, and assumptions, where legitimate changes must be made.

12.5 USE CASE ANALYSIS

12.5.1 IDM for Autonomous Driving Vehicle

IDM is vital for the successful adoption of new technologies through innovation such as autonomous driving vehicles. This involves long-term planning and resource allocation, focusing on market positioning, competitive analysis, technological investments, and sustainability initiatives (Paden et al., 2016). Professionals can follow sequential steps in analysing critical situations to identify alternatives before making decisions. The following steps can be addressed for the identified use case scenario.

Market opportunity: Manufacturers need to identify their target market segments, whether concentrating on luxury electric vehicles, affordable mass-market models, or commercial options. Gaining insights into demographic preferences, geographic trends, and consumer behaviour is crucial for effectively positioning electric cars. Strategic decisions regarding branding, marketing initiatives, and distribution channels can greatly influence market penetration and brand loyalty.

Competitive analysis: The autonomous vehicle market is marked by intense competition among established automakers, new entrants, and technology companies. Companies seeking to differentiate their products must comprehensively analyse competitors' strengths, weaknesses, market share, and innovation strategies. Strategic decision-making should prioritize the identification of unique selling propositions, such as cutting-edge battery technology or superior range, to effectively stand out in a crowded marketplace.

Technological investments: The swift evolution of technology within the autonomous driving vehicle sector demands strategic R&D decisions. Companies must carefully consider their investment areas, such as battery technology, autonomous driving capabilities, and charging infrastructure. Partnering with technology firms, startups, and research institutions can foster innovation and improve operational efficiency.

Sustainability initiatives: As stakeholders increasingly champion sustainable practices, strategic decision-making in autonomous vehicle development must incorporate environmental considerations. This includes assessing the carbon footprint of supply chains, ensuring the responsible sourcing of raw materials, and committing to battery recycling at the end of their lifecycle. Companies that prioritize sustainability bolster their brand reputation and adhere to regulatory requirements.

IDM fosters creativity and develops novel solutions to complex challenges (Carmody-Bubb, 2023). The autonomous vehicle industry relies on innovation, as technological advancements and shifting consumer demands necessitate continuous adaptation and ingenuity.

- Organizations should foster a culture of innovation by encouraging risk-taking and creative thinking. Empowering employees to suggest new ideas, experiment with technologies, and collaborate across departments is essential. Cross-functional teams that blend engineering, user experience design and research, marketing, and customer insights can improve IDM during development.
- Utilizing data-driven insights: Data analytics is essential for promoting innovation in the current information-rich landscape. Companies can extract

insights from consumer behaviour, market trends, and vehicle performance to inform design and functionality decisions. Using data-driven methodologies enables evidence-based decision-making, reducing uncertainty and ensuring alignment between development goals and market needs.

- Customer-centric design: Engaging customers helps identify pain points and preferences, which inform innovative features. Gathering user feedback and conducting focus groups allows developers to understand how consumers interact with electric vehicles and what improvements they want. Prioritizing the end-user experience – whether in range, charging solutions, or technology can significantly impact product success.
- Open innovation: Collaboration with external partners, including technology firms, academic institutions, and research organizations, can foster groundbreaking innovations in electric vehicle development. Open innovation facilitates knowledge exchange and enables companies to leverage new ideas, technologies, and business models. This collaborative approach may lead to advancements in battery performance, charging speed, or the integration of smart technologies within vehicles.

IDM is crucial for autonomous vehicle development but poses challenges such as regulatory changes and market dynamics. Companies must balance short-term goals with long-term vision in a fast-paced environment. Focusing on continuous market research and adaptable strategies can enhance resilience and promote growth in the electric vehicle market.

Creating a dynamic decision-making framework based on agile principles for autonomous driving car development enables teams to quickly adapt to technological changes, consumer demands, and market fluctuations. Agile methodologies emphasize adaptability, iterative progress, and collaboration, which are crucial in the fast-paced automotive industry. Here's an advanced Agile and IDM Framework for autonomous vehicle development:

Define Vision and Objectives

- Establish the vision: Articulate a comprehensive vision for the autonomous vehicle initiative that aligns with strategic organizational objectives, sustainability commitments, and market opportunities.
- Set objectives: Formulate specific, measurable, attainable, relevant, and time-bound goals for the autonomous vehicle and predictive analytics development undertaking.

Create Cross-Functional Teams

- Assemble interdisciplinary teams: To capitalize on varying perspectives and specialized knowledge, construct agile teams comprising diverse experts from engineering, design, marketing, supply chain management, and other relevant fields.
- Empower autonomous decision-making: Delegate authority to teams to make strategic decisions within defined boundaries, fostering ownership and accountability and enhancing motivation and project commitment.

Implement Iterative Development Processes

- Adopt sprint cycles: Define short iterative development phases (sprints) typically lasting 2-4 weeks, allowing for regular re-evaluation of progress and priority adjustments.
- Focus on incremental prototypes: At the end of each sprint, deliver working prototypes, design refinements, or minimum viable products (MVPs) that facilitate ongoing feedback and iterative improvements.

Engage Stakeholders and Customers

- Involve key stakeholders: Actively incorporate input from essential stakeholders such as suppliers, partners, and regulatory bodies into decision-making workflows to gather insights and align strategic expectations.
- Conduct user-centric testing: Utilize rigorous user testing, complemented by surveys and focus groups, to gather feedback from potential consumers and end-users, ensuring the vehicle aligns with market expectations.

Utilize Data-Driven Decision-Making

- Leverage advanced analytics: Implement data analytics and business intelligence tools to evaluate consumer behaviour, market trends, and performance metrics that shape decision-making processes.
- Monitor key performance indicators (KPIs): Define and continuously track crucial KPIs (charging time, battery efficiency, driving range, customer satisfaction) to measure progress and inform strategic decisions.

Collaboration and Communication

- Promote transparent communication: Establish open communication channels within and across teams to exchange insights, challenges, and successes.
- Implement collaboration tools: Utilize project management software and communication platforms that support real-time updates, progress tracking, and constructive feedback mechanisms.

Adapt to Change and Pivot Strategies

- Embrace agile mindsets: Instil a culture that welcomes change and views challenges as opportunities for growth, enabling teams to adapt quickly to emerging trends and consumer insights.
- Conduct retrospectives: Following each sprint or milestone, hold reflection meetings to evaluate successes, identify shortcomings, and refine processes for subsequent iterations.

Risk Management and Continuous Improvement

- Early risk identification: Employ techniques such as SWOT analysis and risk assessment workshops to identify potential risks at the outset of the development cycle.
- Establish mitigation strategies: Develop contingency plans for recognized risks, equipping teams to navigate potential disruptions effectively.

- Encourage a culture of continuous improvement: Promote ongoing learning and process optimization, motivating teams to share insights and adapt methodologies to enhance operational efficiency.

The agile and IDM framework for autonomous vehicle development highlights the need for flexibility, collaboration, and responsiveness in a changing environment. By focusing on iterative progress, stakeholder engagement, and data-driven strategies, organizations can effectively tackle the complexities of autonomous vehicle development. This approach fosters innovation and quality in products that meet the rising demand for autonomous vehicles while supporting sustainability and innovation principles essential for success in the sector.

12.5.2 IDM for Oil Refinery

In high-risk industries like energy or chemical manufacturing, poor decisions can lead to loss of life, environmental harm, or massive financial damage. Studies show that around 80% of major accidents are linked to human error rather than technical failure. This highlights the need for decision processes that are structured, responsive, and effective under pressure.

This task applies three IDMs to the 2022 BP-Husky Toledo Refinery explosion, a $597 million incident that revealed serious flaws in safety management and judgement (CSB, 2023). Our goal is to evaluate how each model can help prevent similar failures by improving decision quality at different organizational levels.

Prior to the incident, audits repeatedly flagged mechanical integrity issues – outdated equipment, faulty valves, and missing safety shut-offs. Although technical solutions existed, poor decision-making and lack of leadership delayed action. The failure was not technical but managerial: a reluctance to prioritize risk. The hypothetical scenario assumes a task force formed three months earlier to draft and execute a safety overhaul, tasked with assessing multiple strategic options.

In high-risk industrial environments such as petroleum refineries, safety incidents can escalate rapidly, requiring decision-makers to respond with agility, innovation, and precision. Operators face an alarm flood situation with thousands of simultaneous alerts, increasing cognitive overload and making it difficult to identify and act on the most critical warnings. The refinery's existing safeguards are heavily reliant on human intervention without sufficient engineered controls or clear procedures for abnormal situations.

In this context, a fast, adaptive, and IDM approach is essential – not only to prevent accidents but also to manage ongoing crises effectively. This scenario demands the integration of real-time process data, automated monitoring systems, human factors considerations, and adaptive feedback mechanisms. Decision-makers must quickly synthesize information, prioritize risks, and implement effective controls while navigating uncertainty and pressure.

Moreover, the model should support continuous learning by incorporating incident outcomes back into the decision process, enabling refinements in safety protocols and operational procedures. Experimentation with advanced technologies like AI for alarm management, predictive analytics for hazard detection, and digital simulations to assess responses in virtual environments are vital components of this scenario, fostering a culture of innovation and resilience.

To effectively address such a complex and rapidly evolving emergency, the innovation IDM, enhanced by design thinking and AI-augmented decision support, was selected as the best framework. This model facilitates iterative problem-solving grounded in empathy, experimentation, and continuous learning while harnessing AI's ability to analyse large volumes of real-time data.

As emphasized in judgemental and rational decision-making, modern decision environments in high-risk industries are shaped by big data analytics, machine learning, and adaptive feedback systems – all crucial to managing volatile, uncertain, complex, and ambiguous (VUCA) situations such as the BP Toledo incident. The incident's complexity, characterized by unexpected process deviations, alarm floods, and human cognitive overload, underscores the limitations of traditional static protocols.

The IDM, supported by AI-driven simulations and alarm prioritization, supports rapid hypothesis generation, scenario evaluation, and decision iteration. Design thinking adds value by centring operator experience, fostering collaborative ideation of solutions, and promoting prototyping of new safeguards and control interfaces.

This model aligns with the operational reality of refineries facing dynamic hazards, where timely decisions are critical and must be refined continually through feedback loops. It empowers teams to adapt quickly, experiment safely, and learn from each event, thereby strengthening safety culture and operational resilience. The application of the agile and IDM model in the BP Toledo Refinery context involves the following detailed steps:

Identify: Sophisticated sensors and control systems continuously monitor critical parameters such as vessel level, pressure, and flow. In the incident scenario, AI-enabled alarm management would identify the abnormal rise of liquid naphtha in a vapour-only vessel by filtering through thousands of concurrent alarms to highlight the critical anomaly. This early detection reduces operator cognitive burden and directs attention to the root cause.

Search: Upon detection, a multidisciplinary rapid response team comprising operations, safety, engineering, and data analytics experts convenes – either virtually or physically. Using real-time dashboards and digital twins of refinery processes, the team explores potential sources of the upset, including valve misalignment and upstream process conditions. AI tools leverage historical data from prior incidents and near misses to suggest failure modes and mitigation tactics.

Evaluate: The team evaluates responses, such as isolating the affected vessel, adjusting upstream flows, or deploying emergency shutdowns. Simulations predict how each action impacts process stability and safety risks. AI models estimate the likelihood of vapour cloud formation and ignition under various scenarios, factoring in human workload and environmental conditions to optimize decisions.

Select: After weighing risks, benefits, and feasibility, the team selects a course of action integrating automated controls and human interventions. Empowering operators with stop-work authority ensures frontline decision-making is respected. The decision balances operational continuity with maximal safety, preventing further escalation.

Implement: The selected response is executed through coordinated efforts between automated systems (e.g., valve actuations, alarm adjustments) and operator actions (e.g., controlled depressurization). Continuous monitoring verifies response effectiveness, with AI adapting parameters dynamically to process feedback and prevent secondary incidents.

Feedback and learn: Following incident stabilization, detailed data and decision logs feed into root cause analyses and model refinement. Safety teams update operating procedures, alarm management philosophies, and training programs based on lessons learned. AI models are retrained with new data, improving predictive accuracy and supporting safer future operations.

Inputs: The inputs to the agile and IDM model are both diverse and sophisticated, reflecting the complexity of modern refinery operations. At the core are real-time operational data streams, which include high-frequency sensor readings monitoring critical variables such as pressure, temperature, fluid levels, and flow rates throughout the refinery process units. These data are continuously fed into advanced alarm management systems that prioritize alerts to reduce operator overload. Additionally, AI models play a vital role by analysing historical data, predicting potential failures, and generating risk scores that inform decision-making. Cross-functional teams composed of operators, safety engineers, data scientists, and process experts provide the human insight necessary to interpret these data within the operational context. This integration of technological and human inputs enables a more comprehensive understanding of evolving situations, leading to faster and more informed decisions.

Assumptions: The successful application of this model rests on several key assumptions. First, it presumes that the data feeding into AI and decision-support systems are accurate, reliable, and timely. Any significant data quality issues or sensor failures could severely undermine model effectiveness. Second, the model assumes that the refinery's systems and infrastructure possess sufficient flexibility and automation capabilities to implement rapid changes in process conditions or control settings. Without this flexibility, the agility inherent in the model would be constrained. Third, the approach depends on teams being trained, prepared, and willing to engage in rapid, iterative decision cycles. This requires a culture that values adaptability, open communication, and the willingness to challenge assumptions and experiment with innovative solutions under pressure. It also assumes that operators have clear authority and trust to act decisively based on emerging information and AI recommendations.

In fast-changing situations like the alarm floods, IDM/ADM stood out. These models focus on speed, feedback, and action. Operators used AI to detect issues, dashboards to assess responses, and design-thinking tools to act fast and smart. What matters here is responsiveness. IDM gives teams more autonomy and encourages quick learning through trial and error. It is ideal for emergencies, where delays could be costly, and where flexibility is more useful than consensus.

12.6 CONCLUSION

This chapter explains IDM and extends analysis through use case scenarios that show how clear, structured models and innovative decisions are vital in high-risk environments. When complex and critical systems are involved, relying on habits or guesswork can lead to profound consequences. The chapter highlights that no single model is enough for making rational decisions in complex scenarios. IDM brings expert input and consensus, helps leaders navigate complexity, and IDM models support quick, adaptive responses. Going forward, the model should be part of everyday operations and leadership training to help teams stay prepared and respond more safely under pressure.

The innovative and agile decision-making explains and explores valuable insights for business continuity, and it can apply a few methods such as Roger's model, OODA Loop, and Double-Loop Learning method, which are effective and efficient ways to identify innovative solutions for complex situations. Then, the chapter explains IDM for autonomous driving vehicle development, and the oil refinery process like energy or chemical manufacturing, in complex adaptive systems.

An agile and IDM framework for autonomous vehicle development underscores the importance of flexibility, collaboration, and rapid responsiveness in a continuously evolving environment. By emphasizing iterative progress, active stakeholder engagement, and data-driven strategies, organizations can better manage the technical, regulatory, and ethical complexities inherent in autonomous systems. Such an approach not only fosters innovation and product quality to meet growing market demand but also aligns with sustainability and long-term innovation objectives critical to sector success. Within this context, fast and adaptive decision-making becomes essential – not only to prevent accidents, but also to manage emerging and ongoing crises effectively. This requires the integration of real-time operational data, automated monitoring technologies, human factors analysis, and continuous feedback mechanisms, enabling decision-makers to rapidly synthesize information, prioritize risks, and implement effective controls under conditions of uncertainty and pressure.

BIBLIOGRAPHY

Ahearne, J. F. (1986). *Three mile Island and Bhopal: Lessons learned and not learned.* Hazards: Technology and Fairness, National Academy Press.

Argyris, C. (2002). Double-loop learning, teaching, and research. *Academy of Management Learning & Education*, 1(2), 206–218. https://doi.org/10.5465/amle.2002.8509400.

Argyris, C., & Schön, D. A. (1978). *Organizational learning: A theory of action perspective.* Addison-Wesley Pub. Co.

Balakrishnan, J., & Dwivedi, Y. K. (2021). Role of cognitive absorption in building user trust and experience. *Psychology & Marketing*, 38(4), 643–668. https://doi.org/10.1002/mar.21462.

Baron, J. (2007). *Thinking and deciding* (4th ed.). Cambridge Univ. Press.

Boyd, J. (1996). *The essence of winning and losing*. Unpublished briefing. Retrieved from the Project White Horse archive. www.projectwhitehorse.com/pdfs/essence_of_winning_losing.pdf

Brown, T. (2009). *Change by design: How design thinking creates new alternatives for business and society.* Harvard Business Press.

Carmody-Bubb, M. (2023). *Cognition and decision making in complex adaptive systems: The human factor in organizational performance* (1st ed.). Springer. https://doi.org/10.1007/978-3-031-31929-7.

Fu, J. S., Shumate, M., & Contractor, N. (2020). Organizational and individual innovation decisions in an interorganizational system: Social influence and decision-making authority. *Journal of Communication*, 70(4), 497–521.

Gerstein, M., Ellsberg, D., & Ellsberg, M. (2009). Flirting with disaster: Why accidents are rarely accidental. *Disaster Prevention and Management*, 18(2), 209–209. https://doi.org/10.1108/dpm.2009.18.2.209.3.

Gigerenzer, G. (2014). *Risk savvy: How to make good decisions*. Allen Lane.

Goleman, D., Boyatzis, R. E., McKee, A., & Finkelstein, S. (2015). *HBR's 10 must reads on emotional intelligence (with featured article" what makes a leader?" by Daniel Goleman)(HBR's 10 must reads)*. Harvard Business Review Press.

Harrison, F. L., & Lock, D. (2004). *Advanced project management: A structured approach* (4th ed. by Frederick Harrison and Dennis Lock.). Gower.

Heifetz, R. A., Grashow, A., & Linsky, M. (2009). *The practice of adaptive leadership: Tools and tactics for changing your organization and the world*. Harvard Business P.

Huang, M.-H., & Rust, R. T. (2022). A framework for collaborative artificial intelligence in marketing. *Journal of Retailing*, 98(2), 209–223. https://doi.org/10.1016/j.jretai.2021.03.001.

Kahneman, D., Lovallo, D., & Sibony, O. (2011). Before you make that big decision. *Harvard Business Review*, 89(6), 50–60.

Klein, G. Sources of Power. (1999). *How people make decisions*. MIT Press.

Krishnan, C., Mariappan, J., Abraham, A., Gaur, L., Abraham, A., & Gaur, L. (2024). The AI revolution in E-commerce: Personalization and predictive analytics. In L. Gaur & A. Abraham (Eds.), *Role of explainable artificial intelligence in E-commerce* (Vol. 1094, pp. 53–64). Springer. https://doi.org/10.1007/978-3-031-55615-9_4.

NRC Fact Sheet. (2018). *Backgrounder on the three mile Island accident*. United States Nuclear Regulatory Commission.

Paden, B., Čáp, M., Yong, S. Z., Yershov, D., & Frazzoli, E. (2016). A survey of motion planning and control techniques for self-driving urban vehicles. *IEEE Transactions on Intelligent Vehicles,* 1(1), 33–55. https://doi.org/10.1109/TIV.2016.2578706

Perrow, C. (1999). *Normal accidents: Living with high-risk technologies*. Princeton University Press.

Pfrang, E. O., & Marshall, R. (1982). Collapse of the Kansas city hyatt regency walkways. *Civil Engineering—ASCE,* 52(7), 65–69.

Post, C., Lokshin, B., & Boone, C. (2021). Adding women to the C-suite changes how companies think. *Harvard Business Review*, 6(April), 1–7.

Rogers, E. M. (2003). *Diffusion of innovations* (5th ed.). Free Press. ISBN: 978-0743222099.

Shahid, N. U., & Sheikh, N. J. (2021). Impact of big data on innovation, competitive advantage, productivity, and decision making: Literature review. *Open Journal of Business and Management*, 9(02), 586.

Simon, H. A. (1979). Rational decision making in business organizations. *The American Economic Review*, 69(4), 493–513.

U.S. Chemical Safety and Hazard Investigation Board (CSB). (2023). *Fatal explosion and fire at the BP-Husky Toledo Refinery* (Investigation Report No. 2022-01-I-OH). CSB. www.csb.gov/bp-husky-toledo-refinery-fatal-explosion-and-fire/

Weick, K. E. (1988). Enacted sensemaking in crisis situations. *Journal of Management Studies*, 25(4), 305–317. https://doi.org/10.1111/j.1467-6486.1988.tb00039.x.

Index

For Product Safety Concerns and Information please contact our EU representative GPSR@taylorandfrancis.com Taylor & Francis Verlag GmbH, Kaufingerstraße 24, 80331 München, Germany

Batch number: 10739608

Printed by Printforce, the Netherlands